Understanding Coronavirus

Since the identification of the first cases of the coronavirus in December 2019 in Wuhan, China, there has been a significant amount of confusion regarding the origin and spread of the so-called "coronavirus," officially named SARS-CoV-2, and the cause of the disease COVID-19.

Conflicting messages from the media and officials across different countries and organizations, the abundance of disparate sources of information, unfounded conspiracy theories on the origins of the newly emerging virus, and the inconsistent public health measures across different countries have all served to increase the level of anxiety in the population.

Where did the virus come from? How is it transmitted? How does it cause disease? Is it like flu? What is a pandemic? What can we do to stop its spread? Written by a leading expert, this concise and accessible introduction provides answers to the most common questions surrounding coronavirus for a general audience.

Raul Rabadan is a Professor in the Departments of Systems Biology and Biomedical Informatics at Columbia University. He is the Director of the Program for Mathematical Genomics at Columbia University. From 2001 to 2003, Dr. Rabadan was a fellow at the Theoretical Physics Division at CERN, the European Organization for Nuclear Research, in Geneva, Switzerland. In 2003 he joined the School of Natural Sciences at the Institute for Advanced Study in Princeton, New Jersey. He has been faculty at Columbia University since 2008. He has been named one of *Popular Science*'s Brilliant 10 (2010), a Stewart Trust Fellow (2013), and he received the Harold and Golden Lamport (2014), Diz Pintado (2018) and Phillip Sharp (2018) awards. He is the co-author with Andrew J. Blumberg, of *Topological Data Analysis for Genomics and Evolution* (2020). Dr. Rabadan's current interest focuses on uncovering patterns of evolution in biological systems – in particular, RNA viruses and cancer.

The *Understanding Life* Series is for anyone wanting an engaging and concise way into a key biological topic. Offering a multi-disciplinary perspective, these accessible guides address common misconceptions and misunderstandings in a thoughtful way to help stimulate debate and encourage a more in-depth understanding. Written by leading thinkers in each field, these books are for anyone wanting an expert overview that will enable a deeper understanding of each topic.

Series Editor: Kostas Kampourakis (http://kampourakis.com/)

Forthcoming titles:

Understanding Coronavirus

RAUL RABADAN
Columbia University, New York

CAMBRIDGE
UNIVERSITY PRESS

CAMBRIDGE
UNIVERSITY PRESS

University Printing House, Cambridge CB2 8BS, United Kingdom

One Liberty Plaza, 20th Floor, New York, NY 10006, USA

477 Williamstown Road, Port Melbourne, VIC 3207, Australia

314–321, 3rd Floor, Plot 3, Splendor Forum, Jasola District Centre,
New Delhi – 110025, India

79 Anson Road, #06-04/06, Singapore 079906

Cambridge University Press is part of the University of Cambridge.

It furthers the University's mission by disseminating knowledge in the pursuit of
education, learning, and research at the highest international levels of excellence.

www.cambridge.org
Information on this title: www.cambridge.org/9781108826716
DOI: 10.1017/9781108920254

First published 2020

A catalogue record for this publication is available from the British Library.

Library of Congress Cataloging-in-Publication Data
Names: Rabadan, Raul, author.
Title: Understanding coronavirus / Raul Rabadan, Columbia University.
Description: Cambridge, United Kingdom; New York, NY: Cambridge University Press, 2020. |
    Series: Understanding life | Includes bibliographical references and index.
Identifiers: LCCN 2020017915 (print) | LCCN 2020017916 (ebook) | ISBN 9781108826716
    (paperback) | ISBN 9781108920254 (epub)
Subjects: LCSH: COVID-19 (Disease)
Classification: LCC RA644.C67 R33 2020 (print) | LCC RA644.C67 (ebook) | DDC 616.2/41–dc23
LC record available at https://lccn.loc.gov/2020017915
LC ebook record available at https://lccn.loc.gov/2020017916

ISBN 978-1-108-82671-6 Paperback

"An extremely concise and important book that everyone should read to understand the dynamics of the COVID-19 pandemic."

Siddhartha Mukherjee, *Columbia University and author of* The Gene *and* The Emperor of All Maladies

"With all the technology and medical knowledge of the twenty-first century, a pandemic virus has defeated us. This book tells us why and how that could happen and what we can do about it. All this wrapped up in a clear, understandable, and interesting way. We learn what is happening to us now and how to better prepare for the future."

Arnold J. Levine, *Institute for Advanced Study, New Jersey*

"… an essential book for the first pandemic of the twenty-first century, COVID-19 caused by SARS-CoV-2 … a wonderfully concise and accessible explanation of everything you want to know about the virus, the disease, and the outbreak. I highly recommend this book."

Vincent Racaniello, *Columbia University*

This book is dedicated to my parents, Fernando and Felicitas, and to the healthcare workers who are fighting for all of us.

Contents

Foreword by Series Editor

Understanding Coronavirus, by Raul Rabadan, is not only a timely book given the current pandemic and the concerns about this virus worldwide; it is also a very useful book that explains the current situation and provides a solid understanding of SARS-CoV-2 in particular, and of viruses and pandemics more broadly. It covers the epidemiology and the biology of this virus, as well as relates it to the SARS and influenza viruses and outbreaks of the past. The book begins with the broader picture about how viruses spread, and then zooms in on what coronaviruses are and how they compare to other viruses, providing the necessary context for understanding the COVID-19 pandemic. The author is a well-known expert, who is closely monitoring current developments, and here provides the interested reader with a well-written, engaging, and up-to-date book.

Kostas Kampourakis, Series Editor

Preface

On a Friday evening in March 2020 I received a phone call from an old friend of mine, Luca. Luca lives in Treviso, a beautiful old town in the north of Italy, close to Venice. He was worried about my family, knowing that the number of coronavirus cases and deaths in New York was increasing at an alarming rate. I could sense that his voice was weak, with a mixture of exhaustion and anxiety. "I feel better now, but it has been a terrible month here," he told me. "For me," he paused, "it all started with a very high fever and a terrible cough a few days later. The cough felt like something was burning inside my lungs." Luca was feeling better at this point, but the description of the situation was not encouraging. "Streets are completely deserted in Treviso and in Venice. You never would have thought about that." That was some time ago, and Luca's account is now just one of many, including from family and friends. By April, New York, the city where I live, was a desert of empty streets, closed restaurants, and few pedestrians. People in their homes followed the development of events with obsessive persistence, tracking the terrifying figures and official announcements. We are all learning and trying to adjust to the "new normal": social distancing, quarantines, telecommuting, home schooling, to name a few.

The identification of the first cases of the coronavirus began in December 2019 in Wuhan, in the Chinese province of Hubei. Since then, there has been a significant amount of confusion regarding the origin and the spread of the coronavirus, officially named SARS-CoV-2, and the severity of the disease, named COVID-19. This confusion has been amplified by conflicting messages from the media, disparate communications from officials across different countries and organizations, and by the radically different measures taken in different parts of the world, ranging from locking down entire regions to a mild denial of the situation. Making things even more disconcerting, various media sources equated this epidemic to mild seasonal flu, the 1918 "Spanish flu" pandemic, or the 2003 SARS outbreak. Each of those outbreaks was caused by a separate virus in very disparate historical moments with very different outcomes. The lack of a clear message, the profusion of conflicting sources and comparisons, the unfounded conspiracy theories on the origins of the newly emerging virus, and the dissonant public health measures across different countries increased the levels of anxiety in the population.

The idea for this book was born through conversations with Cambridge University Press commissioning editor Katrina Halliday and her series editor, Kostas Kampourakis, regarding the current need for simple explanations clarifying some of the confusion generated since the beginning of the pandemic. It aims to provide a concise introduction to the COVID-19 coronavirus through a set of basic questions about the virus and the disease. The book is designed to inform a general reader, someone with an interest in learning more about the coronavirus without having to go to the scientific literature. Topics include the basic molecular biology and epidemiology of the virus, a bit of genomics, a description of the origin and evolution of the virus, and a comparison to other respiratory viruses. I also provide some conceptual tools to help frame the questions and answers addressed in the book.

This book was not planned in advance. I was supposed to enjoy a sabbatical year traveling and meeting colleagues around the world. That did not happen. With conferences cancelled and travel bans in place, it became evident that I was going to enjoy some quiet time, at home with my family. Worried by the rapid evolution of circumstances, I was drawn into research on the virus and the disease, together with colleagues at Columbia University. The sabbatical and frustrated travel plans allowed me to undertake the challenge of writing this book.

I would like to thank my home institution, Columbia University, as well as the Institute for Advanced Study in Princeton, and the Center for Theoretical Physics at Columbia University in New York City for hosting me during the writing of this book. I would like to thank Paula Ralph-Birkett, Andrew Chen, and Suzanne Christen for the initial editing of the book, and Zixuan Wang for the figures. Thanks also to Jean-Michel Bertoli, Katrina Halliday, Kostas

Kampourakis, Mathew Kleban, and Cristina Rabadan, who provided constant feedback on content, structure, and ideas for the book. Arnold Levine carefully read the book and gave insightful comments and ideas. I would like to mention, in particular, Ioan Filip and Juan Patino Galindo for interesting collaborations on coronavirus genomics. I have discussed many aspects of the new coronavirus with members of my laboratory at Columbia University, including Luis Aparicio, Francesco Brundu, Mathieu Carriere, Oliver Elliott, Karen Gomez, Zhaoqi Liu, Tomin Perea-Chamblee, Wesley Tansey, Anqi Wang, and Junfei Zhao. I would like also to acknowledge friends and colleagues for useful feedback and comments, including Gyan Bhanot, Julian Bruat, Reuben Danzing, Bernard Dayrit, Lam Hui, Martin Hyatt, Hossein Khiabanian, Luca Magri, Carmen McConnell, Do-Hyun Nam, Massimo Porrati, Leonardo Rastelli, Jeffrey Shaman, Andrea Severin, and Jiguang Wang.

Most of all, I would like to thank my family, Jean-Michel, Emma and Alex, for their constant patience and support while I was writing this book.

Abbreviations

ACE2	angiotensin-converting enzyme 2
AIDS	Acquired immunodeficiency syndrome
CCoV	canine coronavirus
COVID-19	Coronavirus disease 19
FCoV	feline coronavirus
HIV	human immunodeficiency virus
ICTV	International Committee on Taxonomy of Viruses
ICU	intensive care unit
ILI	influenza-like illnesses
MERS-CoV	Middle East respiratory syndrome
PCR	polymerase chain reaction
RBD	receptor binding domain
R_0	basic reproductive number
SARS	Severe acute respiratory syndrome
WHO	World Health Organization

1 Introduction

The single biggest threat to man's continued dominance on the planet is the virus.

Joshua Lederberg

Viruses populate the world between the living and the non-living, the molecules that can duplicate themselves and the ones that cannot. Inherent in the organization and properties of viruses are many of the secrets of life ...

Arnold Levine

At the end of December 2019, an outbreak of pneumonia cases of unknown origin was reported in Wuhan, Hubei province, China. The patients presented with high fever and had difficulty breathing. Most of these cases were related to the Huanan Seafood Wholesale Market, where, in addition to seafood, a variety of live animals were also sold. Other infections occurred in people staying at a nearby hotel on December 23–27. All tests carried out by the Chinese Center for Disease Control and Prevention for known viruses and bacteria were negative, indicating the presence of a previously unreported agent. A new virus was isolated and its genome sequenced, revealing a similarity with SARS-like coronaviruses found in bats. Although very similar to the virus causing severe acute respiratory syndrome (SARS) in 2003, it was different enough to be considered a new human-infecting coronavirus. Clusters of infected families, together with transmission in medical settings, indicated that the virus had the ability to undergo human-to-human transmission. A month later, by the beginning of February 2020, the virus was found in several countries across the globe, and on March 11, 2020, the World Health Organization (WHO) declared it a global pandemic. The disease caused by the new coronavirus was called coronavirus disease 19, or COVID-19.

The rapid pace of these events led to significant confusion. Attitudes and perceptions in the population varied dramatically, from denial to serious concern and panic, mimicking the disparate comments and actions taken by public authorities and the media. After the declaration of the pandemic and the first serious outbreaks in Wuhan, Northern Italy, Spain, and Iran, it became clear that the emergence of this virus is a serious threat, and can lead to a significant overloading of healthcare systems. By the end of March 2020, the USA, the UK, India, and most countries in Europe had reported an escalating number of cases and deaths, and had implemented extensive public health measures, including lockdowns. The associated effects on the economy are daunting, including international travel bans, market uncertainty, and significant reduction of demand for and production of goods, among many others.

Unfortunately, the confusion of the first few months of the pandemic has led to a profusion of myths, large amounts of inconsequential information, and conspiracy theories that have infected the Internet faster than the virus has spread around the world. In trying to make sense of the situation, and to create a coherent narrative that incorporates the overwhelming data, many questions have arisen: questions about the nature of the virus and the disease it causes, about its changes, and about the future. This book addresses some of these questions. I have decided to structure the book in the form of a dialogue, of simple questions and answers. Most of these questions came from family, friends, and colleagues.

This book is aimed at the lay reader, one who has minimal knowledge of biology, virology, epidemiology, or medicine in general. I have tried to make the chapters self-contained, and they can be read in any order, although I recommend reading the first four chapters first, in order to get a clearer understanding of the biological and epidemiological concepts that are discussed in the chapters about specific viruses and outbreaks. Because this book is a short introduction to the topic, there are some important details that are overlooked. To compensate for the superficiality in how some themes have been treated, I have included at the end of the book a list of recommended reading material that will guide the interested reader to more in-depth treatment of specific topics. This material has been selected from among recent scientific journal papers of broader scope, and from textbooks. I would encourage the enthusiastic reader to follow up with these references. I apologize to some of the researchers whose work I have not been able to discuss or mention due to the introductory nature of this book.

Viruses are fascinating entities that awaken our deepest fears. The history of humankind is literally plagued with the narration of the devastating effects of infectious diseases, in which viruses have been major players. Smallpox killed one in every three people it infected, with an estimated 300 million deaths in the past century. The infamous Spanish Influenza of 1918 shocked the world with its rapid spread, completely overwhelming healthcare systems, and with its vicious attack on the young adult population. The human immunodeficiency virus (HIV) in the 1980s marked a then-young generation and challenged a rapidly evolving society. Rotavirus infection, a vaccine-preventable disease, is one of the most common causes of diarrhea in young children, and kills more than 100,000 children every year. Many other examples, recent and historical, easily come to mind.

Once an infectious pathogen appears, we would like to understand and quantify how it is spreading, what its effects are in the population, and how the efficacy of different public health measures can be evaluated. In the rapid expansion of the COVID-19 virus around the world, we have observed and experienced the role of drastic public health policies that have changed our social lives dramatically, and we have witnessed the rapid growth of cases and deaths associated with the disease. Chapter 2 deals with basic concepts in epidemiology – the science of evaluating the distribution of diseases and different control measures.

What do we know about the virus that causes COVID-19? The coronavirus disease, or COVID-19, is caused by the SARS coronavirus 2, or SARS-CoV-2. Chapter 3 explores viruses, and coronaviruses in particular. Viruses are the most common biological entity on Earth and are present in every realm of the surface of this planet. Only a very small fraction of them interact with humans, and only a small fraction of those are pathogenic. The pathogenic viruses, however, have captured most of the attention of the scientific community. Coronaviruses constitute a particular type of virus that can be found in mammals and birds. Some coronaviruses cause disease in humans, but most of them infect other species, such as bats, without apparent disease. Four coronavirus types are found commonly in humans and induce typical cold symptoms. Others can cause severe disease, like bronchitis in chickens or diarrhea in pigs. Some, as we have seen with the virus causing COVID-19, can cause severe disease in humans. Many questions come to mind. Is this a new virus? Where is it coming from? How does it relate to other coronaviruses? In the third chapter of this book I provide some basic notions of what viruses are, and describe coronaviruses in particular. I explain the different types of coronaviruses and where they can be found. All coronaviruses share a common but highly distinctive structure. I also briefly explain how they enter and leave infected cells.

How was the coronavirus that causes COVID-19 able to infect and spread in humans? To answer this question, we need to understand how viruses evolve. Viruses are the tiniest and most rapidly evolving biological entities known. Changes in viral genomes happen almost continually. All changes in viruses can be read in their tiny genomes, which keep all the information on the virus and its history. Reading genomes is like reading a history book, where the main characters are viruses. This record not only portrays the history, but also allows one to infer the rules of the processes that dictate the changes. Through the recent developments in genomic technologies, viral genomes can be sequenced rapidly and their changes can be observed almost in real time. As the COVID-19-causing virus spreads across the globe, we will be able to follow a parallel history by reading the genomes of the viruses collected in different parts of the world.

Chapter 4 explains the two main mutational mechanisms that drive the evolution of coronaviruses. The first one is what is known as the "sloppiness" of the replication machinery. Once a virus infects a cell, it makes tens of thousands of copies of itself. But these copies are sometimes (often) imperfect, with small variations on the main theme. Many times, these changes lead to a faulty copy. But sometimes, the new virus can acquire new abilities that become useful to the virus, such as the ability to enter a new type of cell or to evade recognition by the immune system of the organism it is infecting. But an even more dramatic mechanism is pervasive in coronaviruses: recombination. In a recombination event, two different viruses can swap genomic material rapidly, quickly acquiring new abilities. The combination of these two processes – sloppiness and recombination – shapes the evolution of coronaviruses. We will talk about these two mechanisms and how they can be read from viral genomes.

Chapters 2–4 provide a background to contextualize the emergence of the virus that causes COVID-19, which is discussed in Chapter 5. Using genomic information, we relate the genome of the new virus, SARS-CoV-2, to other known viruses and where they were found. The new virus is related to SARS-CoV, the agent that caused the 2002–2003 SARS outbreak, and to many other viruses found in other species, mostly bats. I narrate the first known events in the history of this outbreak, how it was first identified, and how it has been evolving. I then discuss the disease caused by the virus – COVID-19 – its symptoms, and how it causes disease and death. I also devote some time to the demographics of the populations at risk, how it is affecting more men than women, and the effects on children.

Chapter 6 is devoted to the outbreak that occurred in 2002 and 2003 due to one of the closest relatives to the virus that causes COVID-19. That outbreak was SARS, and the virus was the SARS coronavirus, a close relative of SARS-CoV-2. These are the only two viruses in the same virus species that are known to have caused outbreaks in humans. There are remarkable similarities between the SARS outbreak of 2002–2003 and the COVID-19 outbreak in 2019–2020. The two viruses have many similarities in their genes, in the type of cells they infect, in their way of entering cells, and in how they interact with the cell machinery and immune system. It is not surprising that the diseases caused by these two viruses share certain similarities. More interestingly, we can learn many things about the new virus causing COVID-19 from the work that scientists have carried out with the virus causing SARS. The basic biology and the clinically acquired knowledge from related viruses can help to accelerate the discovery of potential treatments for COVID-19.

Chapter 7 is a scientific misfit. It is about a virus, but not a coronavirus. Rather, it is about a virus that has been used widely as a comparison: influenza. The elements for comparison are obvious. Influenza causes respiratory diseases; it spreads through surfaces and air droplets in coughs and sneezes; and it causes severe disease in the elderly. These are all elements that are shared with COVID-19. But, in many other aspects, the SARS-CoV-2 and the influenza viruses are very different, and the diseases caused by them, and the severity of those diseases, are very different. Most importantly, for seasonal influenza, there is at least partial immunity in the population, and we have vaccines and specific drugs for treatment. None of this is true for COVID-19. The lack of immunity to the COVID-19-causing virus has taken an immunologically unprotected population by surprise, leading to a dramatic surge in cases that has pushed healthcare systems to the verge of collapse. This rapid surprise attack has occurred in the past in the context of pandemic influenza, most notably in the infamous Spanish Influenza of 1918. In 1918, it was not known that the disease was caused by a virus, and part of the world was still embroiled in a devastating war. That virus, however, was not a coronavirus, and the diseases, the populations most affected, and the healthcare systems were very different. It is, however, instructive to compare some of the historical events of the Spanish Influenza of 1918 to the COVID-19 pandemic of 2020, such as how different places dealt with the unmanageable surge in the number of cases.

The last chapter of this book is about testing, and prospects for therapies and vaccines. At the time of writing there are neither specific drugs that can dramatically reduce mortality nor vaccines – but there are many ideas. Many of these ideas trace back to the 2002–2003 SARS outbreak, and because the outbreak was controlled by July 2003, they were not tested in clinical settings. The rapid evolution of COVID-19 has accelerated the testing of many of these approaches, and has also led to the generation and rapid implementation of innovative ideas for therapies and vaccines.

2 How Is the Coronavirus Spreading?

It is not if an epidemic will occur, but when!

Robert Shope

Once an outbreak starts, it is important to quantify how a disease is spreading, how it is affecting the population, and how different public health measures will have an impact on its effects. Epidemiology is the study of how a disease is distributed in a population, and of the different factors that determine this distribution. These studies can help to quantify the main population factors that led to the introduction and spread of an infectious disease in the population and the conditions that are associated with the severity of the disease. Epidemiology can also be used to assess the current extent of the disease and the effectiveness of different interventions, including different therapies and public health measures. Finally, it can also help to make predictions on likely future scenarios, given the current assessment of the situation and the different measures taken.

In the first few months of the COVID-19 pandemic we have witnessed an exponential increase in the number of cases and deaths. The implementation of different containment measures in different parts of the world has led to vastly different outcomes. Whereas in some Asian countries early containment measures were able to stop the infections, in many other places the growing number of cases made containment unfeasible, forcing a change to mitigation strategies to reduce the impact on the healthcare system and the number of fatalities associated with lack of adequate care (Figure 2.1). For the current COVID-19 outbreak, we would like to quantify what conditions enabled the emergence of the disease, how fast the disease is spreading in the human population, how deadly it is, the different factors that affect disease severity,

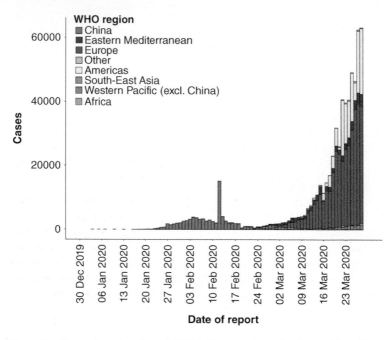

Figure 2.1 **The number of confirmed COVID-19 cases reported to the World Health Organization (WHO) from the beginning of the pandemic at the end of December 2019 until March 29, 2020 by different regions of the world.** The green bars represent the cases reported to the WHO at the beginning of the pandemic, with the vast majority of cases from Wuhan in Hubei province in China. The outbreak in China was rapidly controlled by imposing a lockdown of the region and strong confinement measures. The total number of deaths in Hubei attributed to the virus during this period was estimated to be over 3000. The number of cases in other regions of the world remain very low until the beginning of March 2020, when an exponentially growing number of infections were identified in Europe and the USA. The number of deaths in Europe and the USA follow suit.

how effective the current public health measures are, and the possible future scenarios of the pandemic.

How Long Does It Take to Have Symptoms?

The incubation period is the time from infection to the onset of the disease – that is, the appearance of the first symptoms. This is an important concept, because this period can vary according to the disease and the person. During the incubation period, there are no symptoms, and it is hard to assess whether a person is infectious or will become sick.

A well-known measure against infectious diseases is so-called quarantine. Where does this term come from? In the 1300s, Europe was ravaged by the plague. In order to prevent the disease from entering a city, all boats were required to anchor for 40 days. These 40 days were considered a safe time to assess whether anyone aboard presented the symptoms of the plague. These 40 days, or *quaranta giorni*, is where the word quarantine comes from. A person will become ill 2–6 days after being infected with the bubonic plague, but it takes almost 40 days from infection to death.

The incubation period varies dramatically across different infectious diseases. The incubation period for influenza, for instance, is a couple of days. For the common cold, it is about four days. However, the incubation period for rabies in humans can be 1–2 months. In the COVID-19 outbreak the incubation period has been estimated to be five days, similar to SARS. However, a few rare cases have been reported in which this incubation period has been more than two weeks. Measures of isolation, as with the original quarantine, should take into consideration these longer incubation periods.

How Many People Will the Virus Infect?

The attack rate is the number of people developing a disease or infected with an infectious agent divided by the total of number of people susceptible to the disease. These rates are usually estimated from all cases from the beginning to the end of the epidemic. In an ongoing epidemic, the rate cannot be measured, but it can be estimated through modeling. In the case of a new infectious agent in the population, such as SARS-CoV-2, everyone is susceptible, and the attack rate is the fraction of the population that has been infected with the virus. The attack rate will depend on many factors, including how quickly the infectious disease spreads, and what measures can be taken to avoid its spread.

In the first influenza pandemic of the twentieth century, the Spanish Influenza, it was estimated that one-third of the global population was infected with the virus. In an ongoing outbreak, it is hard to estimate the fraction of the population that will become infected. It will depend on how fast the infectious agent spreads and how effective the measures taken to control its spread are.

How Quickly Does the Virus Spread in the Population?

One way of quantifying the infectiousness of a pathogen is the basic reproduction number (R_0), which is the estimated number of new infections per infected individual. If this number is smaller than 1, meaning that a person rarely infects another person, the number of infected individuals in the population will decrease with time, until the infectious agent will not be able to propagate (Figure 2.2).

But if the basic reproduction number is greater than 1, the number of infections will grow exponentially. For instance, imagine that an infectious agent has a basic reproduction number of 2. This means that a single infected individual will infect, on average, another two people (Figure 2.3). These two newly infected people could then infect another two each, resulting in four newly infected people. And so on. If these new infections occur five days after the previous infection, then in a 30-day month, there are six five-day periods, and a single person can lead to $2 \times 2 \times 2 \times 2 \times 2 \times 2 = 64$ new infections – and in two months that's 4096 new infections. By six months, the entire world population will have been infected. This type of growth is called exponential

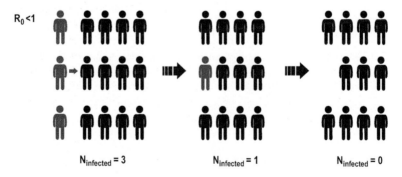

$R_0 < 1$

$N_{infected} = 3$ $N_{infected} = 1$ $N_{infected} = 0$

Figure 2.2 A representation of the evolution of an infectious agent with low basic reproduction number or R_0. A way to estimate how fast an infectious disease is spreading is to estimate the mean number of infections caused by a single infected individual. This illustration represents a case in which a few infected individuals (three in this case, represented in red) interact with uninfected individuals (in black). If the basic reproduction number is smaller than 1, the number of new infections diminishes. Eventually the infectious agent will disappear from the population.

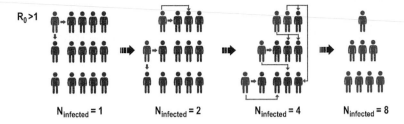

Figure 2.3 **A representation of the evolution of an infectious agent with high basic reproduction number or R_0.** We present a case in which an infectious agent has an estimated R_0 bigger than 1. In such a case, a single infected person can infect several others, and in successive infections this number can increase exponentially. R_0 depends very much on the infectious agent. The R_0 of SARS-CoV-2 is estimated to be 2–3, indicating that the number of infective cases will increase exponentially in the first months of the outbreak.

and characterizes the initial growth of infections in an uncontrolled setting. Under exponential growth, the number of cases can increase very quickly (Figure 2.4). Exponential growth nicely illustrates the idea that early interventions are important. For instance, if by social distancing we can reduce the mean number of contacts by 50%, the number of new infections will be reduced to only one. In the first case, thousands of new infections could saturate a hospital, while in the second they can be monitored and contained. This is an example of how social distancing and social responsibility can generate large effects in the course of an infection. Of course, the growth of these numbers will depend on the number of individuals who are susceptible, and the contact between infected and non-infected individuals. Strong quarantine measures can reduce the basic reproductive number to less than 1.

This R_0 number varies dramatically across different infectious agents. Some viruses transmit poorly, like the Middle East respiratory syndrome coronavirus (MERS-CoV), showing a R_0 of less than 1. This explains why all MERS outbreaks are small, usually happening in hospitals. For seasonal influenza, this number has been estimated to be slightly larger than 1, and the R_0 of the 1918 Spanish Influenza was close to 2. For SARS-CoV and SARS-CoV-2 this number is about 3, which means that every infected person is able to infect another three people. Other infectious diseases can have a much larger basic reproduction number, such as polio and measles.

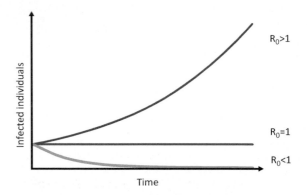

Figure 2.4 The exponential growth of the number of infections depends on the basic reproduction number or R_0. Cases with $R_0 > 1$ will see an exponential increase in the number of infections in a completely susceptible population, if no containment measures are implemented. If the basic reproduction number is smaller than 1, the number of cases will decrease exponentially. A basic reproduction number of 1 will keep the same number of infected cases steady over time as every infected person will infect one other person on average.

How Many People Will Die from the Disease?

The *infection fatality rate* is the number of individuals dying from the disease in a particular period of time divided by the total number of people infected. But sometimes the total number of infected individuals is unknown. This number can be approximated by the *case fatality rate*, or the number of fatalities over the total number of reported cases. If all the infected individuals have been reported, the *infection fatality* and the *case fatality rate* will be the same. But in diseases with unreported or untested cases, as with COVID-19, the two numbers could be very different.

The *mortality rate* of a disease is the fraction of the population who die from the disease in a particular period of time. The main difference between infection fatality rate and mortality rate is that the infection fatality rate is the fraction of the population who die from the disease compared with the number of individuals infected, whereas the mortality rate is compared to the total population. A concrete example is presented in Figure 2.5.

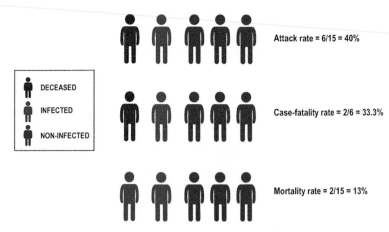

Figure 2.5 An illustration of three relevant measures to characterize the extent and mortality of an infectious disease. Imagine a population of 15 individuals, of whom 6 individuals became infected with disease and 2 of those died from the disease. The attack rate is the fraction of the population that contracts the disease; in this case it is 6 of the 15 members of the population, so the attack rate is 40%. The infection fatality rate is the fraction of the infected cases who die from the disease, in this case it is 2/6, or 33%. The total mortality rate is the fraction of the population that died from the disease, so 2 people out of the 15, or 13%. A localized outbreak could have a very high infection fatality rate, but if the disease does not spread, the mortality rate in the whole population could be low. For instance, the Ebola virus in the outbreak in West Africa in 2014 had a high infection fatality rate of near 40%, whereas the Spanish Influenza was estimated to be close to 10%. However, the total number of deaths in the world from Ebola virus was much smaller (11,000 versus 50 million) because the outbreak remained confined to specific regions.

In the case of the Spanish Influenza of 1918, the number of deaths was estimated to be around 50 million (although estimates range from 20 to 100 million) from a population of 1.8 billion people, giving a mortality rate of 2.7%. If one-third of the population could have been infected, this entails an infection fatality rate close to 10%, although estimates can vary by a factor of 2. Later flu pandemics, in 1957 and 1968, reported fewer deaths at close to two million. Case fatality for other coronaviruses, such as MERS-CoV, are estimated to be much higher, nearly 35%, although the transmissibility is much lower.

Some of the most vulnerable groups to COVID-19 are people of advanced age, and those with other serious diseases or terminal illnesses. If someone is critically ill from a disease, for instance with cancer, and dies after getting infected with COVID-19, should this death be counted as caused by cancer or by COVID-19? In some cases the cause of death is not specified, because the person may not have been tested for the coronavirus. If a person dies from a different cause (e.g. a heart attack) because a coronavirus-overwhelmed healthcare system was not able to provide care, should this death be counted as part of the effects of the virus? One important notion that takes into consideration these cases is the excess mortality – that is, the temporary increase in the mortality rate in a specific population. That can be estimated by counting the number of deaths before and during an outbreak for the same period of time. For instance, the excess mortality during the 1918 Influenza was estimated to be more than 80% – that is, there were 80% more deaths than would have been expected for a similar period of time. There have been a few estimates of the excess mortality in the first weeks of the COVID-19 pandemic (Figure 2.6). For instance, in the Madrid province in Spain in the period March 10–16, 2020, in the first few weeks of the COVID-19 outbreak, 1318 deaths were registered, compared to the expected 794 from estimates in previous weeks. However, only 192 deaths were reported to be associated with COVID-19, suggesting that there could be a significant unreported excess of deaths that may be linked to COVID-19. Similar observations have been made in Wuhan, China and Northern Italy.

How Does the Severity of Diseases Vary?

The severity of most diseases depends on age. For instance, seasonal influenza is a particularly serious disease for infants and the elderly. COVID-19 has been seen to be a particularly serious disease for the elderly. If we want to compare the effect of COVID-19 across different populations, we have to take into account that age distribution varies across countries. Although in Japan the median age is more than 47 years, in many countries in Africa it is below 20. So, if we want to understand the effects of COVID-19 across different populations, we need to take the different population distributions into account.

In COVID-19 the case fatality rates vary dramatically with age. The Chinese Center for Disease Control and Prevention has shown that the case fatality rate under 40 years old was 0.2%, but that it rapidly increases with age to 15%

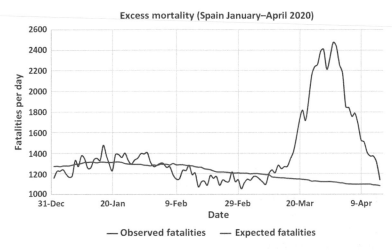

Figure 2.6 Excess mortality, or how to estimate the number of fatalities of COVID-19. Sometimes we do not have information on whether a particular death is related to a cause. For instance, there are many possible causes of a pneumonia-associated death, including different viruses and bacteria. In the case of COVID-19, there have been many suspected deaths that were not tested, for a variety of reasons. When the cause is an infectious agent, such as SARS-CoV-2, we can estimate the number of associated deaths by comparing the number of deaths in the time and region where the agent was circulating with the number of deaths in normal circumstances. In other words, how many more deaths have occurred when an infectious agent has been around in a population, or the excess fatality. This figure shows the number of observed fatalities per day in Spain in the first few months of 2020 (in red) compared to the number of deaths in other years (blue). A dramatic increase in the number of deaths in March was associated with the circulating of SARS-CoV-2 in that country.

above 80 years old. If the case fatality rate age distribution is similar across countries and there are similar numbers of infections, we should expect that the total number of deaths due to COVID-19 in most African countries would be smaller compared to countries with longer life expectancies, such as some countries in Europe, Asia, or North America.

It has also been observed that the severity of infectious diseases can be different in men and women. For seasonal influenza, the rates of hospitalization for males is higher than for females, among both the young and the elderly. The same has been observed in COVID-19 deaths. The difference in the rate of illness (morbidity) and rate of death (mortality) between men and women could be due to a variety of factors, including biological and lifestyle factors. For instance, in many countries smoking is more common in men than women, and smoking is one of the most important risk factors in respiratory complications.

What Is "Flattening the Curve" and Why Is It Important?

During the COVID-19 outbreak we have observed that one of the main factors that has overwhelmed the response to the disease is the rapid surge of cases that need urgent medical attention (Figure 2.7). The surge capacity is the capacity of the healthcare system to accommodate a sudden increase in the number of patients. An overwhelmed healthcare system can occur when a highly infectious disease spreads quickly in a population, and the number of infected people suddenly requiring clinical care is larger than the capacity of the healthcare system. This overwhelmed healthcare system lacks the physical means to provide care to patients and to protect medical personnel, reducing ever further the capacity to respond, and thus accelerating the morbidity and mortality associated with the disease.

The identification of a disease outbreak requires immediate action, with measures including population testing, tracing contacts and isolation, quarantine, and treatment of infected cases. Effective containment measures can reduce the spread and the sudden surge of cases requiring medical attention. If these measures prove ineffective, more drastic measures need to be taken, including closing of public gathering places, schools, theaters, and increased social distancing. More aggressive measures can require lockdowns of entire cities or countries. These measures can "flatten the curve" by lowering the peak number of cases and allowing the healthcare system to respond and provide the clinical care needed in severe cases (Figure 2.7).

Overloading of the capacity of the healthcare system occurred during large pandemics in the past, such as the Spanish Influenza of 1918. During October 1918, 195,000 US citizens died from the virus. In addition, the war in Europe demanded the deployment of nurses to military camps, exacerbating the shortage of medical personnel. There are many historical accounts that record how different cities and countries imposed different measures and how

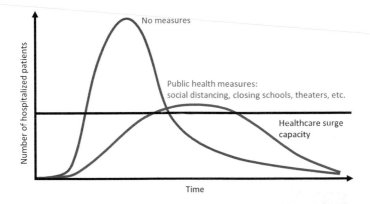

Figure 2.7 Flattening the curve and surge capacity. Overwhelming the healthcare system can reduce the capacity to assist patients and dramatically increase mortality. Early implementation of preventive measures, like social distancing, can mitigate the overwhelming of healthcare systems; delaying the peak gives more time to prepare and to acquire necessary equipment and protection for healthcare workers. Pandemic mitigation strategies include voluntary isolation and quarantine, closure of schools and childcare programs, cancellation of public gatherings, reduction of social interaction at work and increased use of telecommuting, and other social distancing measures. The effect of these mitigation measures can delay the peak and its magnitude, allowing the healthcare system to take care of patients and healthcare workers. In the course of the COVID-19 pandemic, many countries set up field hospitals to accommodate the surge in patient numbers. This figure shows an improvised field hospital in a convention center in the outskirts of Madrid, Spain on April 21st 2020.

effective these measures were. Early aggressive interventions proved to be effective in the reduction of cases and easing of the burden on healthcare systems. We will see in Chapter 8, when talking about the influenza pandemics, a couple of examples of these early interventions. In the case of COVID-19, active testing, tracking, and isolation, together with social distancing, has proven very effective in countries such as South Korea.

What Is Herd Immunity?

An important concept in epidemiology is the idea of herd immunity. When a new infectious disease appears, such as SARS-CoV-2, most of the population is unprotected. This leads to a rapid spread of the infectious disease, with the aforementioned surge in the number of cases and fatalities, and the associated pressures on healthcare systems. However, when a fraction of the population has been exposed previously to the infectious disease, or has been vaccinated, those individuals become immune, slowing the spread of the disease (Figure 2.8). The more people are protected against the disease, the more difficult it is for the infectious agent to spread. If a certain threshold is reached, the infectious agent is unable to spread.

The fraction of the population that needs to become immune in order to stop the spread of the infectious disease is called the *herd immunity threshold*. The more contagious a disease, the higher the herd immunity threshold will be. The herd immunity threshold will depend on the basic reproduction number, R_0. This dependence can be understood easily if we think about the number of potential infections from a single infected individual in a partially immune population: R_0 times the fraction of those susceptible in the total population. If this product is smaller than 1, the infectious agent will not be able to propagate.[1] For the highly infectious measles virus, it has been estimated that 95% of the population needs to be immune. For seasonal influenza, which is less transmissible, the threshold is around 35%.

For COVID-19 the herd immunity threshold has been estimated to be around 50% of the population. That means that we will be able to stop the

---

[1] The herd immunity threshold can be approximated by the formula $f = 1 - 1/R_0$. If R_0 is higher than 2 for SARS-CoV-2, we would expect that more than half of the population need to be immune to control the spread of the virus.

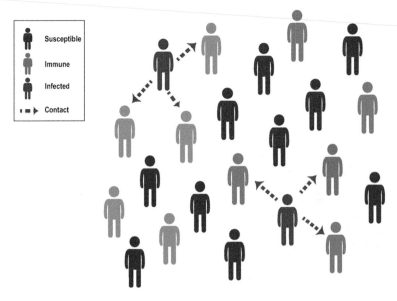

Figure 2.8 Herd immunity. An infectious agent may not be able to propagate in a population in which a significant fraction of individuals is immune to the disease due to previous exposures or vaccination. In order to spread, susceptible individuals (represented in blue) need to be in contact with infectious ones (in red). The more immune individuals (in green) there are in the population the more difficult it will be to spread the infection. The herd immunity threshold is the fraction of the population that needs to be immune in order to preclude disease spread. The herd immunity threshold will depend on the infectiousness of the specific infectious agent. Highly infectious viruses will require a larger fraction of the population to be immune. For instance, the herd immunity threshold for measles is close to 95%. Failure to vaccinate a significant fraction of the population can lead to outbreaks of the disease. For COVID-19, the threshold is lower, probably about 50% of the population.

progression of the virus if more than half of the population has been infected or vaccinated. In the case of a pandemic, such as COVID-19, if containment measures fail, most people will become infected. If they recover, and if they are able to mount a lasting neutralizing immune response, the virus will not be able to re-infect them and the effective rate of transmission will drop. This herd immunity can be dampened if the pathogens evolve to escape the adaptive immune response or if the immune responses are not lasting. Influenza vaccines, for instance, could be ineffective in a fraction of the population, such as the elderly, because the immune system wanes with age. In addition, the virus mutates – this is the reason influenza vaccines need to be updated every year. Herd immunity is one of the main reasons why mass vaccinations have been able to control the spread of many infectious diseases. Lack of compliance with mandatory vaccinations creates larger populations of susceptible individuals, and enables outbreaks of otherwise controllable infectious diseases.

Every virus has its own personality, and coronaviruses are very different from other viruses. To understand how coronaviruses cause disease and how they spread, we need to delve deeper into what coronaviruses are, where they can be found, and how they infect cells. We will look at this in more detail in the next chapter.

3 What Is a Coronavirus?

A virus is a piece of bad news wrapped in protein.

Peter Medawar

We live in a dancing matrix of viruses; they dart, rather like bees, from organism to organism, from plant to insect to mammal to me and back again, and into the sea, tugging along pieces of this genome, strings of genes from that, transplanting grafts of DNA, passing around heredity as though at a great party.

Lewis Thomas

Viruses are amazing creatures. They are the most common, the most diverse, and the fastest-evolving biological entities on Earth. They infect every form of life known, "hijacking" the complex machinery of cells and forcing them into submission. Being much smaller and less complex than cells, they have a unique, tiny kit of "tools" able to regulate the essential elements of cells and to "fool" their defense mechanisms. It should be noted that viruses do not exhibit any of the life properties we usually attribute to cells (such as metabolism, development, or sensitivity) other than reproduction. What viruses practically "do" is to enter cells, their "hosts," and use the cellular machinery to produce new virus particles. It is not surprising that many important discoveries in biology during the last 100 years have been made from, and through, viruses. Viruses have provided fundamental clues to the principles of molecular biology, such as how cells replicate and handle their information and the mechanisms that cause cancers, among many others.

In this chapter, I introduce viruses in general and what coronaviruses are in particular. Viruses are obligate parasites – that is, they need to infect a host in order to replicate. Coronaviruses infect a variety of animals, and humans in particular. But viruses that are able to infect one species are often unable to infect others. Transmission and infection of a new species is possible via a series of processes that I explain in the next chapter, which is about viral evolution. In this chapter I also explain some of the common features among coronaviruses, including the proteins they have and their function, as well as providing an overview of their genome, the container of their genetic information. Reading viral genomes provides useful information about how viruses relate to one another, how they evolve, what their main constituents are, how they replicate, and how they interact with their host. This chapter provides a background for understanding what is special about the SARS-CoV-2 virus, the agent responsible for the COVID-19 pandemic.

What Is a Virus?

Viruses are small, infectious particles that replicate inside cells. They can enter cells and "hijack" the cell's machinery to make thousands of copies of themselves, which in turn can go on to infect other cells. Viruses are very small. The coronaviruses have a diameter of 100 nanometers (0.0000001 meter, or about 1/1000 the width of a hair), and are invisible to standard light microscopes. Human cells are 100 times larger. Whereas human cells typically have tens of thousands of different proteins, the building blocks of all living organisms, viruses can have 10 or so. The small and compact size, and the small number of proteins in viruses, presents a challenge: How can such a small particle effectively "hijack" a complex cell? How can something so small contain the essential ingredients to achieve this? These are some of the most fascinating aspects of viruses, as they can pinpoint the essential aspects of life.

Cells and complex organisms such as humans have adapted to recognize viruses and other pathogens and to eliminate them. They have evolved immune systems that are composed of detection mechanisms that can be "hardwired" in cells (innate immune systems) or learned in the course of responding to infections (adaptive immune systems). The detection mechanisms trigger responses aimed at destroying pathogens and infected cells.

Viruses that do not escape immune recognition are eliminated. Successful viruses, however, manage to escape immune recognition.

Viruses are the most common biological entities on Earth, infecting all cellular organisms, from bacteria to large multicellular organisms like humans. They were identified at the end of the nineteenth century as small particles able to pass through thin-pore porcelain filters (able to filter bacteria and larger cells) and cause disease in tobacco plants. The twentieth century led to the identification of many different viruses associated with common and rare diseases, including measles, rabies, acquired immunodeficiency syndrome (AIDS), smallpox, and influenza, among many others.

Viruses are not only the most common, but also the most diverse biological entities. The genetic information of a virus is encoded in its genome, which provides the specifics of what, when, and in what quantity proteins are produced during infection. It provides information on how the virus replicates and how it escapes the host's immune system. Genomic information in all biological entities is encoded in long molecules – DNA or RNA. Like an audio recording or a book are two different media for storing information, DNA and RNA are two different molecules to encode genetic information. Although all cells use a similar strategy to store genomic information in double-stranded DNA, viruses can use either RNA or DNA (single-stranded or double-stranded). This diversity and the fact that viruses do not share common structures make them difficult to classify. A first broad classification of viruses is by the type of genomic material they use. There are viruses that use DNA as storage material to retain genomic information. Examples of DNA viruses include herpes viruses and smallpox. Other viruses use RNA to store their genetic information; examples of these include the main character of this book, coronaviruses, as well as influenza.

What Are the Coronaviruses?

Coronaviruses are members of a big family of viruses called *Coronaviridae*, which comprises viruses that infect many different mammals and birds. They are called coronaviruses because in electron microscopy images the spikes on their surfaces resemble a crown (corona is the Latin name for crown) (Figure 3.1). The first coronaviruses were identified in the 1930s as infectious agents causing bronchitis.

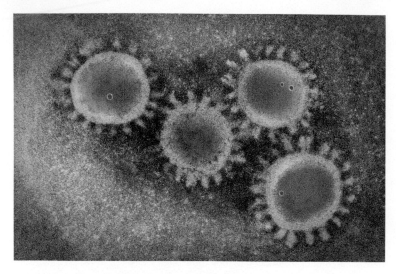

Figure 3.1   Viruses, the most common biological entity on our planet. Viruses are the smallest, the most diverse, and the fastest-evolving biological entities known. Although invisible to ordinary light microscopes, viruses can be observed with electron microscopes. Coronaviruses are characterized by spikes protruding from the viral particle, giving the appearance of a crown (or "corona") shape.

Coronaviruses have a round structure about 100 nanometers (0.0001 millimeters) in diameter. Their genome is the longest of viruses that have RNA as their genetic material, with a genome length of around 30,000 bases (the A, T, C, G letters in DNA [for adenine, thymine, guanine, and cytosine]), and they share a similar genomic organization with related viruses. Most coronaviruses infect only one host, but there are examples, such as the virus causing SARS, that have been shown to infect a variety of mammals, including cats, dogs, Himalayan palm civets, and humans. The diseases associated with coronavirus infections vary in different animals, and include respiratory, enteric, and neurological infections, as well as hepatitis.

Coronaviruses are further classified into different genera based on their genomes:

• *Alphacoronavirus*: These include viruses associated with the common cold in humans, including Human coronavirus 229E and Human coronavirus NL63. These two viruses cause mild respiratory symptoms in humans three days after infection and last for a week; about 30% of infected individuals do not show any symptoms. Lower respiratory tract infections are rare. Other members of this genus have been found in pigs, dogs, and cats.

- *Betacoronavirus*: Members of this genus include SARS-CoV-2, the viruses causing MERS and SARS, and other coronaviruses associated with mild respiratory infections, such as Human coronaviruses HKU1 and OC43. Others have been found in a variety of mammals, including bats, camels, mice, rats, rabbits, hedgehogs, pigs, and cows, among others.
- *Gammacoronavirus* and *Deltacoronavirus*: These have been found infecting different animal species, including birds and pigs.

Coronaviruses infecting humans prior to the 2003 SARS outbreak were considered to cause mild respiratory symptoms. The main character of this book, SARS-CoV-2, belongs to the *Betacoronavirus* genus, and is genetically similar to the virus that caused the SARS outbreak in 2003 (SARS-CoV).

Where Are Coronaviruses Found?

Coronaviruses have been found in a variety of hosts, including birds and mammals. Mammals known to be infected with coronaviruses include bats, camels, dogs, cats, pigs, mask palm civets, and, of course, humans. The diseases vary according to the host and the virus. For instance, the Canine coronavirus (CCoV) is a highly infectious coronavirus causing vomiting and diarrhea in dogs. The Feline coronavirus (FCoV) causes an infectious peritonitis that is frequently lethal in cats.

Viruses circulating in humans can be found in the *Alpha*- and *Betacoronavirus* genera. There are currently seven coronaviruses infecting humans. Four of those (229E, HKU1, NL63, OC43) are continuously circulating in the human population and are associated with mild respiratory infections, although occasionally they can cause pneumonias in babies, the elderly, or people with compromised immune systems. These viruses display a seasonal pattern, being detected during the winter period. Others have been the source of serious outbreaks linked to pneumonias, including SARS-CoV, the agent responsible for the SARS outbreak in 2003, and MERS-CoV, the agent of MERS, which has caused more than 2000 infections since it was first identified in 2012.

What Is the Structure of a Coronavirus?

Coronaviruses share a common structure that is different from other viral families. They are enveloped viruses – that is, they are completely covered by portions of the cell membranes of the infected cells. Cell membranes are the "walls" that separate the interior of the cell from the external environment and help regulate their interactions. They consist of a lipid bilayer and some proteins that mediate the interaction with external factors. Enveloped viruses acquire parts of this cell membrane when budding out of infected cells. Many known viruses have envelopes, including coronaviruses and influenza viruses.

The proteins in a virus can be categorized into two types: the proteins that can be found in the viral particle (the structural proteins); and the ones that are not found within the virus particle, but are expressed when the virus manages to infect the cell (the non-structural proteins).

Figure 3.2 shows a schematic representation of a coronavirus, where the genome is found inside the particle made up of the lipid bilayer membrane and structural proteins. Multiple copies of the nucleocapsid protein (N) bind to the RNA genome of the virus, forming a tubular structure that protects the genome. The membrane (M) protein interacts with the nucleocapsid, and is necessary for packaging the viral RNA. It also plays an important role in assembling the replicating virus in an infected cell.

Among the structural proteins, we should mention the spike protein (S), which is responsible, among other things, for the attachment and entry into the cell. The spike protein protrudes from the envelope of the virus and gives the virus its characteristic "crown" appearance in the electron microscope images (see Figure 3.1). As I describe in this book, the spike protein plays a central role in determining cell and host specificity. The spike binds to proteins in the cell membrane, and if the binding is successful, the cell is infected. The spike can bind to very specific proteins in the host cells, and this ability partially determines the host range of a particular virus.

The non-structural proteins are not found in the viral particle when the virus enters the cell, but rather are produced after infecting it. They are involved in several functions, including replicating the genomic material of the virus. One key protein in viruses is polymerase, the enzyme that drives the replication of the genetic material of the virus. Because coronaviruses have RNA as the main genomic material and cells use DNA, RNA viruses carry the information

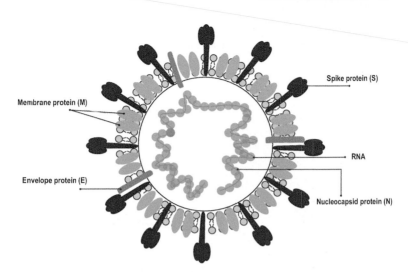

Membrane protein (M)

Spike protein (S)

Envelope protein (E)

RNA

Nucleocapsid protein (N)

Figure 3.2 Coronaviruses share a common structure. The RNA genome resides inside the viral particle bound to the nucleocapsid protein. One of the main functions of the nucleocapsid protein (N) is to protect the genome of the virus by forming a tubular necklace structure of multiple N proteins. The nucleocapsid structure of RNA and protein is then protected by the membrane protein. The membrane protein, the lipid bilayer structure borrowed from the infected cell that gave birth to the virus, has several proteins anchored, including the spike (S), the matrix (M), and the envelope (E). The spike, envelope, matrix, and nucleocapsid are the four structural proteins common to all coronaviruses. The S protein, like most proteins in viruses, has several functions. It is responsible for binding to the cell receptor, the protein mediating the viral entry. It also induces the fusion of the viral and cell membranes. The M protein interacts with the nucleocapsid and triggers virus assembly in infected cells. The E protein is required for the budding of newly formed viruses inside the infected cell.

about the replication machinery that can make copies of RNA from an RNA template, the RNA polymerase. This is able to produce copies of the genome of the virus that will be included in the new viral particles. Other non-structural proteins, the proteases, are in charge of cleaving large viral proteins into small ones. Non-structural proteins have also been found to regulate the innate immune system, a part of the immune system able to detect infections of different kinds, by interacting with proteins in the cell that mediate the recognition and blocking of the subsequent signaling.

The information about all these proteins is encoded in the genome of the virus. Coronaviruses are among the longest RNA viruses known, their genetic material being a single stretch of RNA of near 30,000 nucleotides (the molecules in which the RNA bases A, C, G, U are found – note that, unlike DNA, RNA molecules have a U base [for uracil] instead of T). Genomic technologies allow us to read these genomes, which look like a long sentence of the letters A, C, G, and U. For instance, the genome of one of the first SARS-CoV-2 coronaviruses from Wuhan starts as:

```
AUUAAAGGUUUAUACCUUCCCAGGUAACAAACCAACCAACUUUCGAUCUCUUGUAGAUCUGUUCUCUAAA
CGAACUUUAAAAUCUGUGUGGCUGUCACUCGGCUGCAUGCUUAGUGCACUCACGCAGUAUAAUUAAUAAC
UAAUUACUGUCGUUGACAGGACACGAGUAACUCGUCUAUCUUCUGCAGGCUGCUUACGGUUUCGUCCGUG
UUGCAGCCGAUCAUCAGCACAUCUAGGUUUCGUCCGGGUGUGACCGAAAGGUAAGAUGGAGAGCCUUGUC
CCUGGUUUCAACGAGAAAACACACGUCCAACUCAGUUUGCCUGUUUUACAGGUUCGCGACGUGCUCGUAC
GUGGCUUUGGAGACUCCGUGGAGGAGGUCUUAUCAGAGGCACGUCAACAUCUUAAAGAUGGCACUUGUGG
CUUAGUAGAAGUUGAAAAAGGCGUUUUGCCUCAACUUGAACAGCCCUAUGUGUUCAUCAAACGUUCGGAU
GCUCGAACUGCACCUCAUGGUCAUGUUAUGGUUGAGCUGGUAGCAGAACUCGAAGGCAUUCAGUACGGUC
GUAGUGGUGAGACACUUGGUGUCCUUGUCCCUCAUGUGGGCGAAAUACCAGUGGCUUACCGCAAGGUUCU
UCUUCGUAAGAACGGUAAUAAAGGAGCUGGUGGCCAUAGUUACGGCGCCGAUCUAAAGUCAUUUGACUUA
GGCGACGAGCUUGGCACUGAUCCUUAUGAAGAUUUUCAAGAAAACUGGAACACUAAACAUAGCAGUGGUG
UUACCCGUGAACUCAUGCGUGAGCUUAACGGAGGGGCAUACACUCGCUAUGUCGAUAACAACUUCUGUGG
CCCUGAUGGCUACCCUCUUGAGUGCAUUAAAGACCUUCUAGCACGUGCUGGUAAAGCUUCAUGCACUUUG
UCCGAACAACUGGACUUUAUUGACACUAAGAGGGGUGUAUACUGCUGCCGUGAACAUGAGCAUGAAAUUG
CUUGGUACACGGAACGUUCUGAAAAGAGCUAUGAAUUGCAGACACCUUUUGAAAUUAAAUUGGCAAAGAA
AUUUGACACCUUCAAUGGGGAAUGUCCAAAUUUUGUAUUUCCCUUAAAUUCCAUAAUCAAGACUAUUCAA
CCAAGGGUUGAAAAGAAAAAGCUUGAUGGCUUUAUGGGUAGAAUUCGAUCUGUCUAUCCAGUUGCGUCAC
                    CAAAUGAAUGCAACCAAAUGUGCCUUUCAACU ...
```

The total genome can be written over 10 pages, and it would seem quite incomprehensible. However, we now know how to read the genes coding for the different proteins in the virus – that is, the precise instructions of the components of the virus. Once we have a genome, we can look for the genes, which are the parts of the genome coding for proteins. That can at first be done computationally by searching for long strings of characters within the genome that resemble the distribution of characters we have observed in known genes. The candidate genes can be translated into proteins and compared with proteins found in other viruses. The genome of coronaviruses is split into two main sections. On the first two-thirds of the genome we can find the genes coding for the non-structural proteins – for instance, the RNA polymerase that copies the RNA of the virus. The last part of the genome encodes the information for the structural proteins.

By comparing the genes, and their specific order in the genome, one can assess how similar viruses are. Viruses that are closely related will share a similar genome, a similar ordering of genes, and these genes will encode information about very related proteins. Figure 3.3 is a schematic representation of the genome of three different viruses: SARS-CoV-2, a bat-related SARS-like virus (Bat SL-CoVZC45), and a human virus isolated from the SARS outbreak in 2003 (SARS-CoV). One can appreciate that these three viruses share a very common genomic structure. In dark red are the two larger genes encoding for several proteins that cover the first 22,000 positions of the virus genomes (or 70% of their genome). In green are the remaining 8000 positions of the genome, including the genes coding for the structural proteins (S, E, M, and N), as well as potential genes with less characterized function.

How Does the Coronavirus Enter Cells and Replicate?

Entering the cell is the first step in the process of infection (Figure 3.4). This happens when a protein of the virus binds to a specific protein (receptor) in the membrane of the cell. In coronaviruses, the viral protein that binds to the cell is the spike, encoded in the S gene. The receptor protein depends on the virus and the specific host. One can compare this process to a "key" and a "lock." The virus "key" is the spike protein and the "lock" is the receptor protein of the host cell that can mediate the entry. There are other factors beyond the point of entry that can determine the ability of viruses to infect certain types of cells. Most coronaviruses predominantly infect and replicate in cells in the lungs and intestines.

Once the virus has found the right receptor, the membranes of the cell and the virus fuse. Then the viral genome with the nucleocapsid protein is released into the cytoplasm, the interior of the cell. When the viral RNA reaches the interior of the cell, it is translated into proteins that include the polymerase. Because coronaviruses do not use DNA, they do not use the replication machinery from the infected cell, but their own, an enzyme able to make RNA copies from an RNA template, called RNA-dependent RNA polymerase.

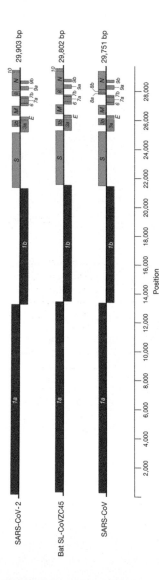

Figure 3.3   Viruses that are related – that is, descend from a common ancestor – are similar at the level of their genome. The genome of the coronaviruses is a strand of approximate 30,000 nucleotides. The genes in the genome encode the information for the viral proteins. Two-thirds of the SARS-CoV-2 genome encode for proteins that are produced upon viral infection, including the proteins for replicating the virus. These non-structural proteins are encoded in the two larger genes that produce two large proteins (polyproteins) that are later cleaved into smaller proteins. The rest of the genes are encoded in the last one-third of the genome (represented in green). Among the genes there, one can find those coding for the structural proteins, the ones that make up the viral particle. These structural proteins include the spike, envelop, membrane, and nucleocapsid. This figure compares the organization of the genes in three related viruses: the virus on the top is a representative of the SARS-CoV-2 viruses, the one in the middle (BAL SL-CoVZC45) is a representative of the SARS-like coronaviruses circulating in bats, and the one at the bottom, SARS-CoV, is one representative of the SARS-CoV, the virus that caused the SARS outbreak in 2003. One can appreciate that the genomes are very similar. Their lengths, marked on the right, vary by 0.5% of the total length. The genes, and the proteins they are coding for, are very similar, and so is the arrangement of these genes in the genome. There are very few differences on some predicted genes, marked in gray, coding for auxiliary proteins whose function in some cases is not well understood. A similar genomic structure is the best way of assessing relatedness in viruses and the main criteria used by the International Committee on Taxonomy of Viruses to decide that the new virus causing COVID-19 is related to the virus that caused SARS in 2003.

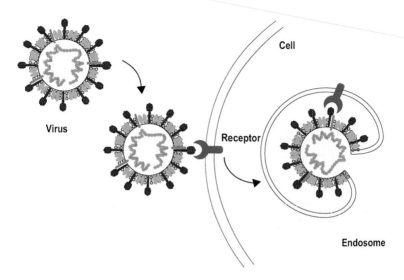

Figure 3.4 The first step of viral entry into a cell is binding to a specific cell protein, the receptor. The first step in a virus infection is the entry into the cell. There are several determinants of tropism, the specific attachment of the virus to specific cells, including host and tissue tropism. One of the main determinants of this tropism is the S protein in the virus. The S protein binds to specific proteins in the surface of the target cells, the receptors. The specific part of the S protein that binds to the receptor is called the receptor binding site (or RBD). The S protein works as a "key" that binds to particular "locks" – the receptors. This key–lock interaction is highly specific: different viruses bind to different receptors. For instance, HCoV-22, a common circulating coronavirus in humans, binds to a protein called APN; whereas SARS-CoV and SARS-CoV-2 bind to ACE2, which is a protein found in the lungs and small intestine, thus determining the tissues that the virus can infect. After the virus has bound to a specific receptor, the membranes of the virus and the cell fuse, forming endosomes, which are membrane-bound compartments that are caused by the inward folding of the membrane.

How Are the New Coronaviruses Released from the Infected Cell?

The new viruses are created when the infected cell starts producing large amounts of replicated viral genomic material and structural proteins. In internal vesicles, the structural proteins interact with each other and the viral genome, assembling new viruses. The vesicles are then released by fusing

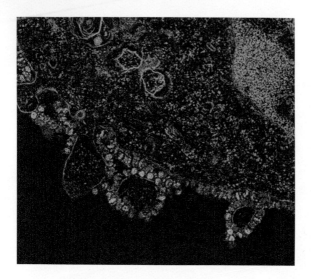

Figure 3.5 Viral release from infected cells. Coronaviruses can produce tens of thousands of viruses from a single infected cell. The process involves the interaction of the nucleocapsid with the RNA genome of the virus and the M protein. The S protein is incorporated, through interactions with the M protein, in the interior of the cell and secreted in vesicles that fuse with the cell membrane. This electron microscope image of SARS-CoV-2 shows coronaviruses budding off an infected cell.

with the membrane from the cells (Figure 3.5). A typical cell infected with a coronavirus can produce tens of thousands of new viruses.

Coronaviruses, like many other viruses, are parasites of cells. They need to enter, use the cellular machinery to replicate, and then be released. At the same time, successful viruses need to avoid being destroyed by the host's immune system. And viruses, like all other organisms, are continuously changing, and they change very quickly. This rapid evolution in viruses determines how viruses can acquire new abilities and infect new hosts, as has happened in the COVID-19 outbreak. But the changing nature of viruses follows rules as they rely on specific mechanisms. In the next chapter, I will describe these mechanisms of viral evolution.

4 How Is the Coronavirus Changing?

Nature is a tinkerer, not an inventor.

Francois Jacob

People have always asked whether evolution is constantly driving onwards and upwards. Is there always going to be improvement? The answer is no: evolution is a progression of form and function, but it is not purposeful.

Sydney Brenner

Theodosius Dobzhansky's essay, "Nothing in Biology Makes Sense Except in the Light of Evolution," reflects on how evolution gives a powerful perspective to biological phenomena, able to integrate disparate pieces of information into coherent narratives. Biology can be messy, with many organisms, cell types, parts, and data. The recent revolution in genomic technology has generated a deluge of data that, correctly interpreted, can illuminate the relationship between and the ancestry of different organisms, the mechanisms that give rise to variability, and how this variability enables organisms to adapt to new environments.

Viruses are not an exception. They keep their genomic information in their core and a significant fraction of their genes and proteins serve to maintain and propagate this genomic information. A virus, like a coronavirus, can create tens of thousands of copies once it infects a cell. These copies, although very similar, present small variations that can affect the ability of the virus to propagate. Viruses that, by chance and due to the large number of these mutations, are able to acquire the ability to infect a new host and successfully establish themselves in a new species become extremely successful in creating multitudes of copies that will continue the expansion of their viral descendants.

Explaining how coronaviruses change, and how they exchange genomic material, is the main goal of this chapter. Coronaviruses evolve in two ways. First, they can accumulate mutations – changes in a single letter of the RNA – that can create small variations in their proteins. But, coronaviruses can also change in a different way. They can exchange genetic information with other coronaviruses in a process called *recombination*. When two viruses enter the same cell, it is possible to generate a new virus with a mosaic genome containing parts from both of them. This process allows a coronavirus to rapidly acquire the ability of another virus by acquiring part of its genomic material.

In this chapter, I explain how viruses change and through what mechanisms. I explain that viruses acquire new variations at great speed compared to cellular organisms such as humans. This rapid evolution can be tracked by reading the genomes of viruses, allowing us to infer how viruses propagate in a population and how they acquire new abilities.

What Is a Mutation?

A mutation is a change in the genome of a biological entity. Mutations are pervasive in viruses. The most common mutations are substitutions or point mutations – the change of one base for another in a particular position of the genome (Figure 4.1). Many of these changes do not affect the virus. Many other mutations can be detrimental, and make the virus less able to compete with other viruses in infecting a new cell. A deleterious change in the spike protein (the point of entry) can preclude the virus from entering a cell, and so it will perish without being able to replicate. A deleterious change in the RNA polymerase can also make a virus unable to replicate, even if it infects a cell. However, and rarely, some mutations can enable a virus to out-compete its relatives. For instance, if a virus gains the ability to elude the detection of the host's immune system, it will be able to replicate in large numbers and most of its progeny will inherit this ability.

····GUGUAGCUGUCACUCAGCUG····

····GUGUAGCUGGCACUCAGCUG····

Figure 4.1 Mutations in genomes. Genomes can be perceived as long texts containing fours letters (A,C,G, and T in the case of DNA; or A,C,G, and U in the case of RNA). The genome of a coronavirus is around 30,000 letters long, whereas the human genome is 3.3 billion letters long. When organisms – including viruses – replicate, they pass this genomic information to their descendants. However, the replication is sometimes faulty, introducing errors that are called mutations. There are many types of mutations. One of the most common ones is when one of the letters changes into another, which is called a substitution or point mutation. In this illustration a U changes to a G. RNA viruses have one of the highest error rates, which can be one change every replication. Human cells have developed an enormous molecular machinery to identify and correct potential mistakes in copying genomic information, which viruses lack.

Viruses, in particular RNA viruses such as the coronaviruses, acquire mutations at a very high rate that accumulate over time.[1] For instance, the estimated number of mutations per year (or evolutionary rate) of the SARS-CoV-2 virus is around 20, or approximately one mutation every couple of weeks. This is very similar to what has been found in other coronaviruses and RNA viruses in general. How these numbers translate into the number of mutations occurring per infection depends on how much time there is between an infection and a transmission from the newly infected person. If, for instance, a person infected with a coronavirus becomes infectious after a week, we can estimate that there will be one mutation every couple of infections. This is an interestingly high rate of mutations that can enable the tracking of the viruses in a community. By comparing a virus with viruses that have been collected in different places around the planet, we could ask what are the likely places of origin of a virus that has emerged in a part of the world. For example, are coronaviruses that have been detected in the USA more related to viruses circulating in Asia or

---

[1] Coronaviruses have a proofreading system that corrects the mistakes made during replication. This proofreading system allows them to have lower mutation rates than other RNA viruses (although much higher than cellular organisms), which could explain why coronaviruses have the longest genomes among RNA viruses.

Europe? These evolutionary inferences could be important for identifying potential sources of outbreaks.

To put this number in perspective we can compare the number of mutations generated during a replication across different organisms (Figure 4.2). Let us take, for example, a coronavirus and a human, and compare the ability to

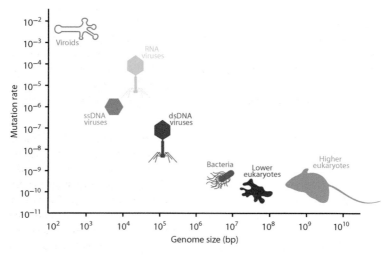

Figure 4.2 **The enormous diversity of life is reflected in the large variability in genome size.** This plot represents in logarithmic scale the variation between the size of a genome (x-axis) and the number of point mutations (per replication per base; y-axis). The shortest entities, the viroids, are small pieces of floating RNA that infect plants. They do not have genes or proteins. Viroid genomes are just a couple hundred bases of RNA. RNA viruses, like the coronaviruses, are the next in size, with a genome length of around 30,000 bases. Coronaviruses have the longest genomes among the RNA viruses. DNA viruses have, like our cells, DNA as genomic material. Their sizes can vary and some recently discovered DNA viruses have genomes of one million bases. The genome sizes of bacteria are also on the million-bases scale. The larger genomes are in eukaryotes, like animals and plants. Our genome length is in the billions. But some plants can have very long genomes that can reach 100 billion bases, like *Paris japonica*, a Japanese flowering plant. Interestingly, the size of the genome is inversely correlated to the mutation rates. Organisms with shorter genomes, like viruses, introduce more errors when replicating their genomes. Large genomes are able to encode error-correcting mechanisms that detect and correct potential mutations that could be deleterious.

generate new variants. They dramatically differ in replication time, the number of offspring, the size of their genomes, and their mutation rates. Coronaviruses produce descendants within a few hours, whereas humans take quite a bit longer, usually 20 or more years. Viruses are able to create tens of thousands of copies, whereas humans produce only a few children. The number of changes in a replication, their mutation rates, are also very different. RNA viruses, such as the coronavirus, can generate a change every few replications, which accumulate every few weeks. The size of the genome of a virus is vastly different from the human genome. A coronavirus genome, the largest genome among the RNA viruses, is 30,000 bases, whereas a human genome has two copies of 3,234,000,000 bases. That is, the human genome is 100,000 times larger than the coronavirus genome. All these differences between viruses and humans provide a glimpse of the complex task that our immune system confronts when dealing with such rapidly evolving pathogens.

This large accumulation of mutations, the large progeny, and the quick replication times make viruses extraordinarily diverse entities, and moving targets for therapeutic developments, including vaccines. Viruses rapidly generate a large number of variations that can quickly adapt to new environments and circumstances. Viruses, like HIV, can escape therapy by quickly acquiring mutations that change the ability of a drug to target them. The continuous genomic changes in viruses like influenza allow them to differentiate and, thus, to evade immune recognition. Vaccines need to be updated frequently to meet these evolutionary challenges.

What Is a Recombination?

In addition to point mutations, viruses can evolve by acquiring genomic material from other viruses and sometimes from the host cells. This is called recombination, and it enables viruses to quickly acquire new abilities. Such phenomena occur extensively in the coronaviruses.[2]

---

[2] Coronaviruses are particularly prone to recombination. This particular mechanism is related to the way the RNA is generated and the jumping of the replication machinery. I recommend the interested reader to check the technical references at the end of the book.

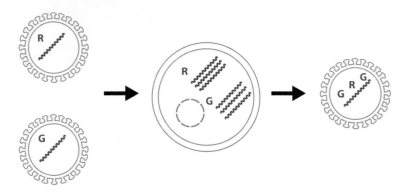

Figure 4.3 **Viruses can rapidly evolve by exchanging genomic material.** When two different viruses co-infect the same cell (in this case, the viruses with the red and the green genome), they can create new viruses whose genome is a combination of the "parental" genomes (in this case the new virus genome contains genomic information from both the red and the green genomes). This phenomenon, called viral recombination, is a pervasive phenomenon in coronaviruses. Recombination allows viruses to rapidly acquire new abilities by combining the "parental" genomes. For instance, a virus can acquire a spike gene that allows the virus to enter a new cell type or a new host.

In particular, if two different coronaviruses infect the same cell simultaneously, they can produce descendants that contain genomic information from both "parental" strains (Figure 4.3). Viruses with different genomes give rise to viral chimeras containing genomic mosaics. For instance, Figure 4.3 shows two viruses with different genomes, one depicted in red and the other in green, co-infecting the same cell. The genomic material of both viruses can mix within the cell, leading to the generation of recombinant genomes (of red and green color).

The mixing, however, is not random. Particular sections of the genome of coronaviruses are prone to be recombinant. Figure 4.4(a) shows the analysis of recombinations that occur in betacoronaviruses. Each of the bars represents a single recombination. The x-axis represents the positions in the genome where new "acquired" material was identified. One of these regions is the one containing the S gene (marked in gray in the figure) coding for the spike protein, which mediates the interaction with the host cell receptor and

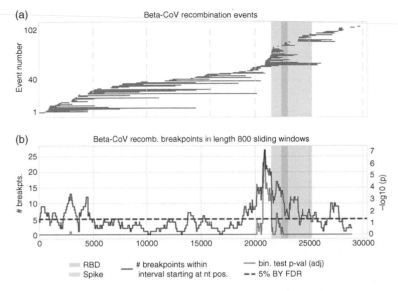

Figure 4.4 Recombinations are very common in betacoronaviruses, the genus of viruses that includes SARS-CoV-2. Recombination events can be inferred by evaluating potential insertions of genomic material. Panel (a) shows a bar for each recombination event spanning a fraction of the genome (x-axis). The recombination can be very small (of a few hundred bases) or large (of tens of thousands of bases). Many of the recombinations are located around the spike gene (in gray). A way of quantifying the density of recombinations is by taking the starting point of each of the segments and evaluating their number along the genome (panel (b), blue line). This plot indicates a clear increase in the number of recombinations near the spike gene.

partially determines the ability of a virus to infect a particular host. That can more easily be appreciated in Figure 4.4(b), which that presents the number of recombinations that occurred at a particular position. The figure shows that recombinations that start near the spike gene are much more frequent than elsewhere in the genome.

Recombination has been hypothesized to be a mechanism for how SARS-CoV-2 acquired the ability to infect human cells. Ancestors of this virus recombined with a SARS-like virus able to enter human cells. Using the key analogy from the previous chapter, through recombination a virus can get the "key" from another virus that enables it to enter a new type of cell.

It is worth reflecting upon the implications of the two mechanisms that we have described. Although point mutations occur rapidly, they can bring about only small changes in the genome. For instance, during a year the genome of a coronavirus can change up to 0.1%. This is a lot, especially compared to humans, where the mean number of changes per year is estimated to be less than 0.000002%. However, recombination is a game changer, as it accelerates evolution by combining biological properties from the different "parental" strains.

How Do We Track Back the Origin of SARS-CoV-2?

The ability to read genomic material using sequencing technologies is allowing the rapid characterization of the genomes of circulating and emerging viruses. This genomic information supports accurate inferences about the characteristics of the virus, and allows a precise comparison with other related viruses.

For example, we might be interested in a newly emergent virus and wish to know what type of virus it is and how it relates to other viruses. We can sequence its genome and see how similar it is to previously reported viruses, and consider in what species these viruses were found. If a new virus, like SARS-CoV-2, is found to be related to viruses found in bats but not to the coronaviruses currently circulating in humans, we can assume an origin from animals (zoonotic origin). Using the genomic information of the new virus and of its relatives, we can also reconstruct the genomes of the ancestors and study the potential changes that enabled the zoonotic transmission.

By comparing related viruses from different locations and different times, we can reconstruct how these viruses may be related to one another. For instance, if we find that the viruses sampled are highly related, we can suspect that there was a single introduction into the community and of subsequent virus evolution during the transmission among the local members of that community. If we find viruses that are not closely related, we can suspect that there were several introductions of these viruses in the community. If we are interested in the origin of an outbreak and the properties of the virus that caused the outbreak, we can use the fact that viruses rapidly accumulate mutations to ascertain when the common ancestor of the sampled viruses appeared. The inference of the genome of a common ancestor can tell us not only about when the outbreak happened, but can also point to the specific alterations that enabled the outbreak.

A common way of capturing the relationships among different genomes and to infer information about ancestors is through a phylogenetic tree. A phylogenetic tree is a representation of the evolutionary history of genomes, which shows the relationships among them and whose branches have a length that represents time. A phylogenetic analysis allows the study of the relationships among different genomes, and makes it possible to reconstruct and infer times of ancestral states. Figure 4.5 shows how a simple phylogenetic tree can capture the relationship among three simple genomes (ACGA, TCGG, and TCGC). The last two genomes, TCGG and TCGC, are very similar, differing in only a single position. They are, therefore, put in two sub-branches in the same branch of a tree and together they form a clade. The internal node represents a common ancestor of these two genomes. This, in turn, is connected to the other more distant genome ACGA via the root, which represents the common ancestor of the three genomes.

The process of inferring phylogenetic relations is widely applied to viral genomes to infer relationships among viruses isolated in different places, times, and hosts. It allows us to infer properties of ancestral states and to estimate times or even places where a common ancestor has occurred. This is particularly useful in the case of outbreaks, as it can help to elucidate the potential origin and the means of transmission.

Viral phylogenetics is also used to track the evolution of the virus across different locations and times. For instance, the H3N2 influenza virus causes seasonal influenza across the world (Figure 4.6). There are many interesting questions regarding how the virus that has circulated in a location, let us say New York this year, relates to the virus that circulated in New York last year, or the virus that has been circulating in Beijing or Paris. Viruses in the same place and season are frequently clustered in related branches on the phylogenetic tree, indicating genetic proximity. In Figure 4.6 we show an example of 1089 H3N2 influenza viruses across

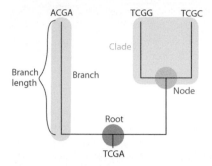

Figure 4.5 Schematic phylogenetic representation of the relationship between three different genomes. This is a simple example to illustrate how the relationship between genomes can be represented by a phylogenetic tree. The three genomes are ACGA, TCGG, and TCGC. Each of the genomes is attached to a leaf of the tree. We then realize that TCGG and TCGC are very similar as they only differ in a base in the last position. We join the two leaves with an internal node representing an ancestral genome, which can be related to the remaining genome by a branch. The three can have a root if we have information about an ancestral state (in this case the state was TCGA). We can also assign to each branch a length, a number that represents the number of differences between states. For instance, the ACGA genome and the root genome TCGA differ in one position. Phylogenetic trees are used pervasively in biology to represent relationships between genomes, species, families, etc. They are frequently used to represent the relationships among the genomes of viruses collected from different times and places. Inferring ancestral states and the relation between different branches provides an idea of how the virus has been expanded in a population or how an emerging virus is related to other viruses previously sampled in other species.

15 seasons. Each season is colored differently. It can be appreciated easily that viruses on the same branch share the same color, suggesting there are many cases where in a single season there were different viruses circulating.

In summary, viruses are extraordinarily fast evolvers. Their evolution rates outcompete all forms of life. The changes in the genomes of coronaviruses are mostly of two types. The first are point mutations or a change of one character in the genome for another. The other, more dramatic, type of change is recombination, the mixing of the genetic material of different viruses to create

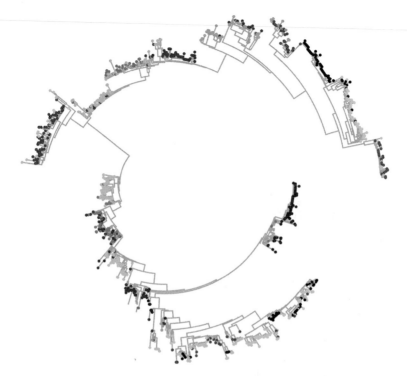

Figure 4.6   Evolution of influenza A virus across different seasons represented in a phylogenetic tree. This tree represents 1089 H3N2 viruses collected in different places in the world and spanning 15 seasons. Each virus is represented by an internal branch of the tree and is colored by the season when it was sampled. It can be appreciated easily that viruses that are genetically related, represented in the same branch of the tree, frequently have the same color. This captures the idea that viruses sampled in the same year are genetically related. But, interestingly, there are branches that show viruses in different colors, indicating that these related viruses spanned several seasons. Also, in some case viruses with the same color appear in different branches of the tree. This indicates that there were several viruses circulating in the season to which that color corresponds. The fact that there could be different viral strains in the same season constitutes a challenge for vaccine design. Every year one H3N2 strain is selected for the vaccine. If in the next season a very different virus or several strains are circulating, the selected vaccine could be inadequate.

a new one. Rapid changes in the genome of viruses allows them to quickly adapt to new environments. How did these changes lead to the appearance of the SARS-CoV-2 coronavirus? Where is the virus coming from and how is it evolving? In the next chapter, I will explain what we know about SARS-CoV-2, the virus causing the disease COVID-19.

5 How Did the COVID-19 Outbreak Start and Evolve?

It's very, very difficult when you have to prepare for something that might not ever happen.

Anthony Fauci

When the first news of the outbreak of a new coronavirus that caused pneumonia in Wuhan appeared in January 2020, it was unclear what effect the virus would have on the population and economy of the world. Despite the repeated advice of health organizations, the containment measures came too late in many places. As cases and associated deaths crept up in different countries, the global concern changed to anxiety. This anxiety hovered between the uncertainty of bland denials and of grim predictions.

In this chapter we will talk about the disease COVID-19 and the virus causing the disease, SARS-CoV-2. I will explain how this virus relates to previously known coronaviruses in humans and other species, recapitulate some of the events in the first month of the then-incipient pandemic, and how the virus spreads and causes disease. I will discuss the morbidity and the mortality of the virus, as well as conditions that worsen disease severity. Finally, I will discuss how this coronavirus spreads and diversifies.

What Is SARS-CoV-2 and What Is COVID-19?

There are three different and related concepts that could result in confusion: the disease, the causative agent of the specific disease, and the species the virus belongs to. The coronavirus disease, or COVID-19, is caused by the severe acute respiratory syndrome coronavirus 2, or SARS-CoV-2. This virus belongs to a viral species, the severe acute respiratory syndrome-related coronaviruses (SARSr-CoV). There are other related viruses within this species.

How are these viruses named? The International Committee on Taxonomy of Viruses (ICTV) is a group of experts who determine how to name and classify viruses based on a series of criteria, including the similarity with other viruses and the hosts they infect. The committee on taxonomy of viruses determined on February 11, 2020 that the new coronavirus responsible for the outbreak in Wuhan in December 2019 belongs to the existing species of SARS-like viruses. This viral species includes the specific coronavirus responsible for the SARS outbreak of 2003, named SARS-CoV. Viral species comprise a group of closely related viruses that often take the name from a founding member – in this case the SARS coronavirus. This similarity is usually based on genomic data: all viruses from SARS-CoV have very closely related genomes. This species of viruses, including SARS-CoV and SARS-CoV-2, belong to the more extensive and more loosely related *Coronaviridae* family (Table 5.1).

On the same day, February 11, 2020, the World Health Organization (WHO), responsible for the name of the disease, announced that COVID-19 was the name of the disease caused by the SARS-CoV-2 virus. Although the virus is similar to SARS-CoV – the viral agent related to SARS – the WHO thought that both clinical and epidemiological data were distinct from SARS. Therefore, they decided to denote the disease as coronavirus disease 2019, or COVID-19.

Where Was SARS-CoV-2 First Reported?

A cluster of similar cases of pneumonia of unknown origin were reported at the end of December 2019 in Wuhan, the capital of the Hubei province in China. Interestingly, most of the cases were reported in relation to the Huanan Seafood Wholesale Market, a market where a large variety of animals are sold, including live animals. Some of the cases were not related to the market, but were located in close proximity, indicating that an infectious agent could be the cause of the mysterious pneumonias. Tests carried out by health authorities for known pathogens, including known viruses and bacteria, were negative, suggesting that the infectious agent was not known. By mid-January, a new virus was isolated and its genome was sequenced. Based on the genomic information, the virus was found to be a betacoronavirus, related

Name of virus	Species	Genus	Family	Name of disease
SARS-CoV-2	Severe acute respiratory syndrome-related coronavirus	*Betacoronavirus*	*Coronaviridae*	Coronavirus disease 19 (COVID-19)
SARS-CoV	Severe acute respiratory syndrome-related coronavirus	*Betacoronavirus*	*Coronaviridae*	Severe acute respiratory syndrome (SARS)
MERS-CoV	Middle East respiratory syndrome-related coronavirus	*Betacoronavirus*	*Coronaviridae*	Middle East respiratory syndrome (MERS)

Table 5.1 Three different coronaviruses that cause severe respiratory syndromes. The three viruses belong to the same genus (and family) of *Betacoronavirus*. SARS-CoV is the infection agent of the severe acute respiratory syndrome (SARS) of 2002–2003; MERS-CoV is the agent responsible of the Middle East respiratory syndrome (MERS) in 2012–2019; and SARS-CoV-2 of COVID-19 in 2019–2020.

to SARS-CoV, the virus that caused the outbreak of SARS in 2003. In the third week of January, cases of the virus were reported in Thailand and Japan. A week later, on January 23, the Chinese government put the city of Wuhan under lockdown, coinciding with the Chinese New Year, which was followed by further measures in Hubei and the whole of China (Figure 5.1). Meanwhile, the number of reported cases increased around the world, leading to similar outbreaks in Italy, Iran, and Spain by mid-March. On March 11, 2020 the WHO declared a global pandemic.

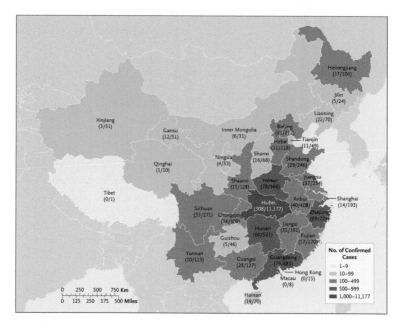

Figure 5.1   COVID-19 cases in China by February 4, 2020. This map shows for each province in China the number of laboratory-confirmed cases of COVID-19 in the official statistics as of February 4, 2020. The epicenter of the outbreak was in the city of Wuhan, in the province of Hubei. The first cluster of cases of the disease was reported in Wuhan in December 2019. Within a month, cases were reported in all provinces of China. A lockdown in the Hubei province, and public health measures across China, led to an effective reduction of the number of cases and deaths by the end of March 2020. Numbers in the map reflect the number of individuals in the study and the total number of officially reported cases by that date.

Looking at the genomes of the viruses circulating in the first few months, one can estimate when the common ancestor to all circulating SARS-CoV-2 viruses occurred. All genomes are very similar, indicating that there is a single common origin to all SARS-CoV-2 and not multiple introductions. One can also deduce that the origin is very recent, estimated to be in mid-November 2019, close to when the first case was identified at the end of November or beginning of December 2019.

The WHO reports daily on newly diagnosed cases and deaths across different regions in the world.

Where Is It Coming From?

The origin of the viruses is still unclear. SARS-like viruses have been sampled in bats and it seems plausible that the ultimate reservoir is found in bat species. Two different scenarios have been proposed to explain how the virus was able to emerge in humans. The first scenario suggests that there could have been some evolution and adaptation in an animal host before it was able to infect humans. As was observed in SARS, other mammals can be infected with SARS-like coronaviruses that use the same entry mechanisms in the same cell type that the viruses infected in humans. The diversity of coronaviruses in bats and other species has not been fully characterized and it is possible that further analysis in different animal species can identify a potential reservoir for the virus. As mentioned above, the first cases were identified in the Huanan market in Wuhan, where several animal species were sold, including rare live animals. The second scenario suggests that some of the changes in the viral genome that enable rapid transmission happened after the virus started to infect humans. In this possible scenario, undetected human-to-human transmission could have led to the adaptation of the virus. This further adaptation could have triggered the detection associated with the clusters of pneumonia cases in Wuhan.

The animal origin and the natural host of SARS-CoV-2 is currently unknown. The closest virus to SARS-CoV-2 currently reported was sequenced by the Wuhan Institute of Virology, which named it RaTG13. RaTG13 was found in an intermediate horseshoe bat (*Rhinolophus affinis*) in caves in Yunnan on July 24, 2013. The horseshoe bat is a common bat species in South and Southeast Asia and central China (Figure 5.2(b)).

(a) (b)

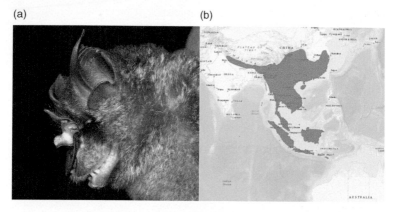

Figure 5.2 SARS-like viruses are found in animals. Bats have been found to host large numbers of viruses, including SARS-like viruses, suggesting that these animals could be the natural reservoir of these viruses. The diversity of viruses in bats also suggests that bats could be a mixing pot for new genetic combinations. This could explain why viruses found in bats in China are similar to the viruses that caused SARS in 2002–2003 and COVID-19. One of the viruses with the closest genome to SARS-CoV-2 was found in an intermediate horseshoe bat (*Rhinolophus affinis*) in South China (a). This species of bat is common in South Asia (b). Other kinds of horseshoe bat have been found to have similar viruses. It has been suspected that the viruses found in bats could infect other mammals, which could then infect humans. How the SARS-CoV-2 managed to find its way to humans is still unclear.

The genomes of these two viruses are 96% identical. Using the current estimates of how these viruses evolve, one can get an idea that the most recent common ancestor to the two viruses existed in the first decade of this century. Sampling of viruses in pangolins has found coronaviruses with small sections of their genome similar to SARS-CoV-2. However, the closest match along the whole genome is, at the moment, the virus found in bats. More systematic sampling of bats and other mammals is likely to identify a potential source of the new virus.

More exhaustive sampling on coronaviruses in different animal species can illuminate the animal sources and the mechanisms of adaptation to humans. Current and banked human samples can also provide information on whether an unreported spread may have occurred.

How Is It Related to Other Coronaviruses?

To get a wider view and perspective on how the SARS-CoV-2 coronavirus relates to other betacoronaviruses, one can compare their genomes. In one of the first analyses of the virus, genomes from the viruses from nine of the first COVID-19 patients who visited Huanan Seafood Wholesale Market in Wuhan were sequenced and compared with related viruses (Figure 5.3). One way of representing this comparison is with a phylogenetic tree. Very similar genomes are represented as nearby leaves. The analysis showed that the new viruses (denoted in Figure 5.3 with an alternative name as 2019-nCoV and colored in red) were all very similar (99.98% identity), indicating that they share a common, very recent origin, compatible with the idea that the first cluster in December 2019 in Wuhan was very close to the most recent common ancestor of these viruses.

The closest relatives in this analysis (RaTG13 was not considered in this tree) were related to some bat SARS-like coronaviruses from Zhoushan, in the Zhejiang province of China. These viruses were found in *Rhinolophus sinicus*, a type of bat of the same genus as the intermediate horseshoe bat, where RaTG13 was isolated. This similitude indicates that the horseshoe bat (*Rhinolophus*) could be a natural source for this virus. These viruses were more distantly related to other viruses found in bats and to SARS-CoV, the causative agent of the 2003 SARS outbreak. As a whole, the analysis showed that SARS-CoV-2 was located in a subgenus of *Betacoronavirus*, related to SARS and bat SARS-like coronaviruses. Overall, the genomes of SARS-CoV-2 and SARS-CoV are 80% identical, although this similarity varies across different regions in their genome.

The phylogenetic reconstruction could be misleading due to recombination – that is, the exchange of genomic material between coronaviruses. Depending on the region of the genome, some viruses can be more similar to each other if there has been an exchange of genomic material between them. One way of detecting recombination is to repeat the phylogenetic analysis across different regions of the viral genome. Inconsistencies in these phylogenies reflect potential recombinations. In particular, when studying the part of the spike gene that binds to the cellular protein of entry (the cell receptor), one finds that SARS-CoV-2 is related to SARS-CoV, the other virus in this group that was known to infect humans (Figure 5.4). The resemblance

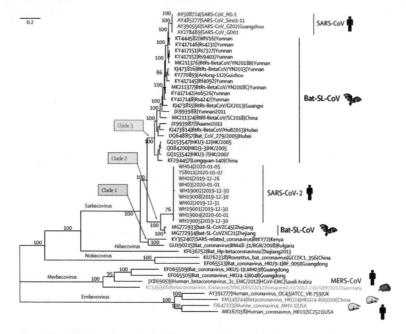

Figure 5.3   Phylogenetic analysis of betacoronaviruses. Phylogenetic trees capture the relationships among related genomes. Genomes are annotated as leaves in the tree. Related viruses appear in clusters. This tree shows the relationships among the genomes of some betacoronaviruses found in a variety of host species: humans, bats, mice, and hedgehogs. Human betacoronaviruses include SARS-CoV-2, annotated as 2019-nCoV, and SARS-CoV, both colored in red. Both SARS-CoV and SARS-CoV-2 are part of *Sarbecovirus*, a subgenus of betacoronaviruses, together with a wide variety of viruses found in bats. SARS-CoV-2 contains nine different genomes from the first month of the pandemic in Wuhan and all appear within a tiny cluster, showing the close similarity of their genomes due to the recent single origin. SARS-CoV viruses collected during the SARS outbreak are also forming a close cluster, indicating an independent single origin of the SARS viruses in 2002. Between them, a large diversity of viruses found in bats suggests that the SARS-CoV and SARS-CoV-2 outbreaks are isolates of zoonotic spillover from bats to humans.

between the only two viruses infecting humans in the port of entry to cells indicates that a potential recombination in the spike gene in one of the ancestors of the new viruses could have led to incorporating the ability to infect humans.

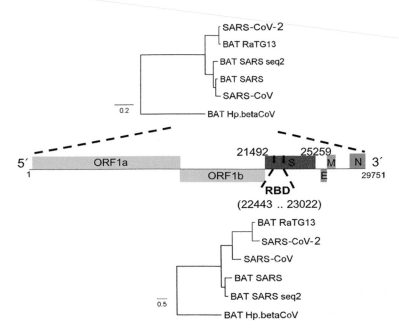

Figure 5.4 **A potential recombination in the ancestors of SARS-CoV-2**. Recombination events are common in coronaviruses, and very frequently they involve the spike gene or sections of it. The recombination could involve the receptor binding domain (RBD), the part of the virus that interacts with the receptor in the cell and contributes to determining the host range of potential infections. Recombinations can be identified by changes in the phylogenetic tree structure if different sections of the genome are taken for the analysis. When we draw a phylogenetic tree with the whole genome, SARS-CoV and SARS-CoV-2 appear on two different branches. But if we restrict our analysis to the RBD, the tree is different and SARS-CoV and SARS-CoV-2 come closer. This suggests a potential recombination in the RBD region, making the receptor binding site of SARS-CoV-2 similar to SARS-CoV. The inference suggests possible scenarios with one recombination at least 11 years ago.

How Does It Enter Human Cells?

Once the virus reaches a new host, it infects cells by attaching to proteins in the cell membrane – the receptor. As we saw in Chapter 2, coronaviruses use the spike protein as a key to enter the cell. Viruses that share the same spike protein can enter the same host cell. In particular, the region of binding in the spike to the receptor is called the receptor binding domain, or RBD. From genetic analysis we have seen that recombinations in the spike make SARS-CoV-2 similar to SARS-CoV in the RBD, suggesting that the viruses share the same receptor.

The SARS-CoV-2 receptor in the cell, the main point of viral entry to the cell, used by SARS-CoV and other human coronaviruses (e.g., HCoV-NL63), has been extensively studied. The receptor is called the angiotensin-converting enzyme 2, or ACE2. The main role of ACE2 is the maturation of a hormone that regulates blood pressure by controlling the constriction of blood vessels and renal function. Altered levels of ACE2 are associated with cardiovascular and renal diseases. The ACE2 protein can be found in many tissues, including the lungs, heart, kidneys, and small intestine (Figure 5.5). ACE2 is expressed in the lower lung, and this expression could be associated with the ability of SARS-CoV-2 to cause pneumonia. The amount of ACE2 protein in many tissues declines with age and also varies between men and women (the gene is located on the X-chromosome). It is unclear whether some of these differences might be contributing to the varied severity of COVID-19 found in different age and sex groups.

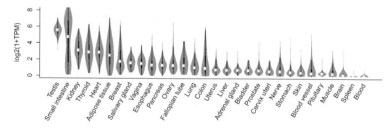

Figure 5.5 Expression of ACE2 across different tissues. Angiotensin-converting enzyme 2, or ACE2, is the receptor, the protein in human cells that SARS-CoV and SARS-CoV-2 use for entering the cells. If this protein were not expressed, the virus could not enter the cell. ACE2 is an important protein whose main role is to regulate the constriction of blood vessels. This figure shows the expression level of ACE2 across different tissues (y-axis). ACE2 is not expressed uniformly, with tissues such as the small intestine, heart, and kidney showing higher levels of ACE2.

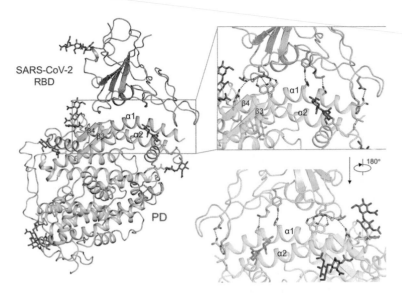

Figure 5.6   Structure of the binding of the virus with the cell receptor. The RBD is the part of the spike gene that interacts with the receptor. The two proteins interact at certain positions that determine the specificity of the interaction. Mutations in these positions are likely to perturb this interaction and reduce the potential of the virus to enter the cell.

Recently, the detailed structure of SARS-CoV RBD and ACE2 has been elucidated in detail using cryogenic electron microscopy, with a resolution of 3 Å (0.0000003 millimeters). This work shows the specific interacting positions in the RBD and the receptor (Figure 5.6). Binding to a different protein or different host requires a change in the protein. The detailed structure of the virus–receptor interaction provides the molecular basis to develop specific therapies that can disrupt or block the entry of the virus to the cell. Interestingly, SARS-CoV-2 and SARS-CoV bind with different affinity to ACE2, potentially contributing to the different infectivity of the two viruses.

How Does It Spread?

SARS-CoV-2 viruses seem to be able to stay in small droplets of water with a half-life of one hour and on metallic or plastic surfaces with a half-life of six hours.

This suggests that the most common form of transmission is through droplets of water (sneezing and coughing) and fomite (clothes, utensils, furniture) transmission. In a droplet, the coronavirus could be in the air for a few seconds before gravity pulls it down. In that period of time someone could come into contact with the droplet and become infected. The virus lifetimes in droplets and on surfaces were found to be similar to SARS-CoV virus. SARS was found to infect many people rapidly in hospitals where patients were treated, and a few cases were reported of super-spreaders, people who infected a large number of contacts. A few cases of super-spreading have been observed for SARS-CoV-2. (Possible explanations for super-spreading are discussed in Chapter 6.)

Clinical studies tracking viral shedding of COVID-19 patients in hospitals have shown that the median duration of viral shedding in these patients is around 20 days, but could be as much as 37 days. It has also been reported that pre-symptomatic cases (infected cases who have not yet developed the disease) and mild cases could be infectious two days before the manifestation of first symptoms.

What Are the Clinical Characteristics?

The initial report from the WHO indicated that 80% of laboratory-confirmed cases have a mild to moderate disease, 14% showed severe disease, and 6% became critical, showing symptoms of respiratory failure and multiple organ dysfunction. Very few people were found to be truly asymptomatic. Further studies have found that asymptomatic infections could be much more common and can significantly contribute to the spread of the virus.

With an incubation period of 5–6 days, mild and moderate disease was observed in a majority of patients who had fever and mild respiratory symptoms, with recovery after a couple of weeks. The most common symptoms in mild cases included fever and dry cough. Diarrhea has been reported in a fraction of patients, indicating intestinal infections. The time to recover in more severe cases was 3–6 weeks from the time of symptom onset. Severe cases have higher viral loads and shed viruses for a longer time, even before the more severe symptoms develop, suggesting that the viral amounts could be used as a marker of prognosis.

Hospitalized cases presented similar scenarios, with difficulty breathing (dyspnea) after a week (Figure 5.7). A study of 191 infected inpatient

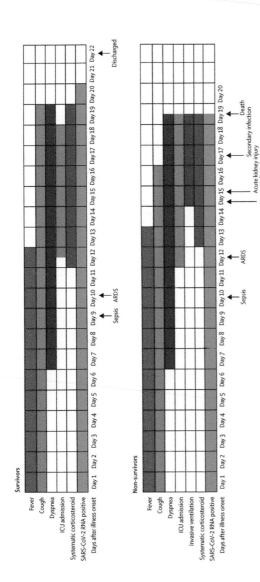

Figure 5.7 **Study of the clinical course of some of the first hospital patients in China, and onset of complications and outcomes.** The study followed 191 hospital patients, of whom 137 were discharged and 54 died in the hospital. The illustration shows the median duration of symptoms and treatments according to outcomes. The most common symptoms in both groups of patients were fever and cough. Six days after the onset of symptoms, some of these patients suffered from shortness of breath (dyspnea) and had to be admitted into the intensive care unit. Some of these patients in day 9 or 10 after the first symptoms had complications such as sepsis or acute respiratory distress syndrome. In the non-survivors, the median time to sepsis was 10 days, to acute respiratory distress syndrome was 12 days, and to acute cardiac injury was 14 days. Most patients needing invasive mechanical ventilation died a few days later. The median time from first symptoms to discharge was 22 days in survivors; in non-survivors the median time to death was 18 days.

adults from hospitals in Wuhan showed that the time to death after the time of first symptoms was 18 days (Figure 5.7). The most common complications were sepsis, respiratory failure, and heart failure. In some severe cases there was an over-activation of the immune system and the lungs became full of fluids, making breathing impossible (acute respiratory distress syndrome [ARDS]).

How Deadly Is It?

The best way of quantifying the deadliness of an emergent pathogen is estimating the number of infected cases that result in death, or the infection fatality rate. Although, in principle, this calculation is just a division of two numbers, it can be extremely tricky for several reasons. First, we often do not know the total number of infected, and it could be hard to estimate the total number of infected cases when only a few cases are reported – usually the more serious conditions. We can distinguish between *case fatality rate* (deaths divided by reported infections) and *infection fatality rate* (deaths divided by inferred number of infections). These numbers could be very different if many infected have not been reported. Unreported infections could be due to not reporting symptoms, showing only mild symptoms, or not presenting the criteria qualifying for the test. The case fatality rate is an overestimate of the infection fatality rate, in particular during the first phases of the outbreak, when only severe cases requiring clinical care are reported.

The second factor is that death occurs some time after the infection. In the case of SARS-CoV-2 infection, it has been estimated to occur within a couple of weeks after detection. This time lag means that during the rapidly spreading phase of the disease, a death can occur weeks after the infection, when the number of cases is much greater than when the infection occurred. Therefore, if we simply divide the number of deaths by the number of cases at that moment in a growing epidemic, we will underestimate the case fatality rate. Other factors that can lead to variability in case fatality rate estimates are the quality of the healthcare system and the particular characteristics of a population. Case fatality rates can increase if adequate care is not provided to critical patients or if the population is older, as could be the case in a nursing home or on a cruise ship.

In the first phases of the SARS-CoV-2 outbreak, the WHO has estimated a case fatality rate of 3.8%, with a significant variability across different locations. Subsequent modeling of infections suggested that a large fraction of all infected cases (86%) in Wuhan until January 23 were not reported, and probably showed milder or no symptoms. If this is confirmed, the infection fatality rate will be much lower, nearer to 0.6%.

Are Age, Sex, and Other Diseases Affecting Death Rates?

It was observed early on that the mortality of COVID-19 increases with age and that more men than women die (4.7% vs. 2.8% of reported infected cases, respectively). The case fatality rate was highest among people over 80 (21.9%), and few deaths have been reported for children or adolescents. Other diseases, also associated with age, are found to play an important role, including hypertension, diabetes, cardiovascular diseases, and chronic respiratory disease, among others. The same pattern has been observed in several countries, including Italy and Spain (Figure 5.8).

The same age and sex biases were observed during the SARS epidemic in 2003, where mortality increased to 50% in individuals over 60 years old. Interestingly, experiments with mice showed the same pattern – young mice were resistant to infection with mouse-adapted SARS viruses, while older mice, mostly male, were more susceptible to infection and death.

There is much speculation about the specific mechanism of age-related severity, including a senescent immune system and mechanisms to dampen inflammation and T-cell responses. There is sexual dimorphism in the immune system – women have higher rates of autoimmune diseases and men have worse outcomes from infections. It has been suggested that females have more active innate immunity, leading to better control of infections. This has been observed not only in humans but many other mammals. Lifestyle factors could also contribute, including the higher rate of cigarette smoking in men in many countries. But at the moment there is no clear explanation of the age or sex profile differences.

Does the Virus Infect Children?

Very few cases of infected children have been found. In the cases in Wuhan, only 2% of cases occurred in youths (below 19 years old), with few cases

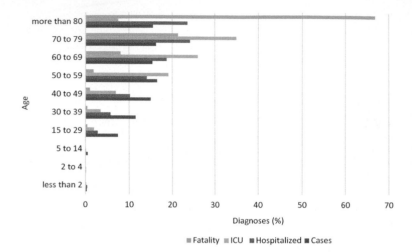

Figure 5.8 **Distribution of diagnosed cases, hospitalized patients, patients in the intensive care unit, and fatalities from COVID-19 as of March 24, 2020 in Spain.** Around the globe, COVID-19 showed a clear age pattern distribution. The fraction of diagnoses were distributed across all age groups, with very few cases among children and young adults. Hospitalization rates steadily increase with age. This dependence on age is even more pronounced in ICU cases. The larger fraction of deaths occurred in the elderly, more than 60% of deaths were in patients who were more than 80 years old. Hospitalization and deaths were more frequent in men than women; the same observations were found in other countries.

developing severe disease. In a retrospective study of 366 hospitalized children with respiratory infections admitted January 7–15, 2020 in the Wuhan area, SARS-CoV-2 was found in only six patients (or 1.6%). The median age of the infected children was three years old, with these patients showing high fever, coughing, and vomiting. Four of them showed pneumonia, and one was admitted into the pediatric intensive care unit (ICU). Epidemiological data on COVID-19 studying 391 cases and 1286 individuals who were in close contact with the cases suggest that children are at similar risk as adults, but they have less severe symptoms.

This pattern mimics what was found in the SARS outbreak in 2003, where no one under 24 years old died, and most of the deaths occurred in

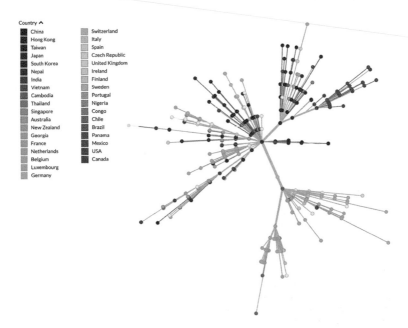

Figure 5.9   Diversification of SARS-CoV-2 as it expands around the world. The genomes of SARS-CoV-2 collected in different locations in the world look very similar, consistent with data that the outbreak started in Wuhan in November 2019. However, RNA viruses mutate very quickly and these mutations introduce a small number of variations that allow comparisons of the genomes of viruses. This tree represents the genetic similarity among SARS-CoV-2 viruses collected in different countries (colored differently here). One can see that viruses cluster and that these clusters are associated with specific regions of the world. For instance, the cluster in the lower part of the figure is mostly green, from viruses sampled in several countries in Europe. Interestingly, there are also some red dots, from samples collected in the USA, that are genetically similar to viruses in Europe, suggesting a travel-related infection. The right-hand cluster on the top right are cases that genetically are very similar, collected from Washington State. The close similarities among these viruses indicate community transmission. The red dots, cases in the USA, appear in different parts of the tree, indicating multiple introductions.

patients over 60. Very few of the SARS-infected cases were children, and only 5% received intensive care. This observation was replicated in mice. Young mice presented similar viral levels as adults, but showed fewer symptoms and mortality. This observation, that the age- or sex-associated morbidity and mortality are similar in mice and humans, suggests a potential biological cause.

How Fast Is the Virus Spreading?

One way of quantifying how fast the virus has been spreading is by the basic reproduction number (R_0), which from the first cases in Wuhan has been estimated to be 2–2.5. This number explains the exponential growth of cases and associated deaths in the population.

Is the Coronavirus Diversifying?

Viruses are constantly evolving, and as the virus has been expanding around the globe more mutations have appeared. These mutations allow tracking of the virus and infections. There are mutations that allow matching related viruses into clades, groups of related viruses that share some common mutations. For instance, viruses circulating in the state of Washington in the USA present a mutation in position 8782 in their genome that viruses circulating in Italy do not have (Figure 5.9). In addition, most viruses circulating in Europe share a mutation in position 3037 that many viruses circulating in Asia do not have. We can use these mutations to pinpoint the origin of the virus infecting a particular individual, if the person has been infected by other members of the community or from a recent introduction and from where.

The changing nature of the virus will create a challenge for therapeutic and vaccine development. For instance, a change in a protein targeted by a vaccine can make the vaccine less effective. The escape mutants can render the therapy ineffective.

The 2019 coronavirus pandemic has been a terrifying experience for a large part of the population of the world. This is not the first outbreak associated with this type of coronavirus. In 2002 and 2003, the SARS coronavirus led to global alarm but fortunately that event was contained. We learned many things about SARS-like coronaviruses then. In the next chapter I will revisit the 2003 SARS outbreak and how it relates to the 2019 pandemic.

6 How Does the COVID-19 Outbreak Compare to the SARS Outbreak in 2003?

When a disease insinuates itself so potently into the imagination of an era, it is often because it impinges on an anxiety latent within that imagination ... SARS set off a panic about global spread and contagion at a time when globalism and social contagion were issues simmering nervously in the West.

Siddhartha Mukherjee

The SARS outbreak provides evidence to support the hypothesis that southern China could be a site for emerging pandemic infectious disease in the future.

Zhong et al., Lancet (2003)

The virus infecting humans that is closest to SARS-CoV-2 is SARS-CoV, the agent that caused the SARS outbreak in 2002 and 2003. These two viruses are very similar in their genomes, in their way of entering cells, and in some of their clinical characteristics. Since 2003, we have learned many things from the virus that caused SARS. We have learned how the virus enters cells, how it replicates, and how it interacts with the immune system. We have learned some of the main factors that contribute to the worsening of the disease. Animal models have been established, and therapies have been developed and proposed. This acquired knowledge can accelerate the discovery of potential treatments for COVID-19.

There are many parallels that can be drawn between the outbreak of SARS in 2002–2003 and the outbreak of COVID-19 in 2019–2020. The two viruses likely started from the same pool of viruses, probably circulating in bats and infecting other species. Both outbreaks were identified as a cluster of pneumonias of unknown origin, and rapid response isolated and identified the viruses within two months. There are similarities in the clinical characteristics of the diseases, such as the incubation period, the mode of propagation, the first symptoms, and the high incidence of lower respiratory tract pneumonias. There are, however, intriguing differences. For instance, SARS was mostly localized in healthcare facilities and family clusters, whereas COVID-19 was able to spread easily beyond these settings. SARS was contained and practically disappeared within a few months, which is not the case for COVID-19. In this chapter, I examine some of these parallels and divergences.

How Did the SARS Outbreak Start?

In November 2002, the first case of an atypical pneumonia was reported in Foshan, near Guangzhou, in the Guangdong province in southern China. Several cases followed in the Guangdong region, including family clusters, suggesting an infectious spread. Many of the patients were transported to hospitals in Guangzhou, the Guangdong capital, for clinical care. This led to many infections within the hospitals there. The disease started with high fever, muscle pain, and respiratory symptoms (dry cough and shortness of breath), developing a few days later into pneumonia. When the WHO was informed in February 2003, the outbreak in Guangdong counted 305 cases and 5 deaths. The cluster included a large number of healthcare workers.

In February 2003, a doctor who treated SARS patients in Guangdong traveled to Hong Kong for a family gathering. He developed the disease while in Hong Kong and died there. Twenty-three other guests staying in the same hotel developed the disease. Many of the infected guests from the hotel traveled to other countries, including Canada, Taiwan, Singapore, and Vietnam, thus spreading the disease. By March 2003, SARS cases were identified in 13 countries, although most cases were reported in China and Hong Kong. The genomes of the viruses isolated in different regions were almost identical, suggesting that all cases were caused by the same agent that had emerged in a short time in the past. The disease was named severe acute respiratory syndrome, or SARS.

In April 2003, the agent causing SARS was isolated and was found to be a coronavirus distantly related to a previously known one (Figure 6.1). The virus was named SARS-CoV, as the SARS-causing coronavirus. None of the healthy controls showed antibody or viral RNA, suggesting that this was a new virus. It was also found that patients did not show immune response to the new virus, suggesting that the virus was new and not cross-reacting with the commonly circulating human coronaviruses 229E and OC43. All these analyses indicated that it was a new virus that the immune systems of the patients had not encountered before.

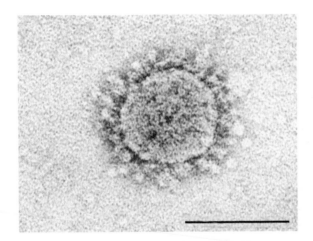

Figure 6.1 Identification of the virus causing SARS in 2003. Electron microscopy image of a SARS-CoV virus showing the typical coronavirus structure with spike projections surrounding the membrane envelope of the virus. The bar on the bottom right-hand side represents a scale of 0.00001 cm (100 nm). The virus was isolated four months after the first few cases were reported in Guangdong in 2002. The genome of the SARS-CoV virus was compared to previously known viral genomes and was found to be a distant relative of other coronaviruses. Very little was known at the time about betacoronaviruses circulating in other species. By contrast, during the 2019 SARS-CoV-2 outbreak, the time taken for the isolation and sequencing of the viral genome was reduced to less than a month after the first cases of the outbreak were reported. This quick action led to the implementation of detection kits and active surveillance and quarantine measures. The outbreak in Wuhan was effectively controlled by March 2020.

Where Was the SARS Virus Coming From?

The SARS virus was not circulating in humans, which triggered a search for the virus in other species. One potential source might be related to some culinary habits in the south of China, where wild game and live animal markets (wet markets) are not uncommon. This hypothesis was reinforced by the fact that some of the early SARS cases in Guangdong included workers at restaurants that featured wild animals on the menu.

In 2003, different studies indicated that the viruses could be isolated from wild animals sold in wet markets in the Guangdong province of China. SARS-like coronaviruses were found in Himalayan palm civets (also called masked palm civets, with the scientific name *Paguma larvata*) and in a raccoon dog in a live-animal market in Shenzhen. Sequencing of the genomes of these viruses indicated that they were very closely related to SARS-CoV, the virus responsible for the SARS outbreak.

Two cases were identified as a waitress and a customer in a restaurant in Guangzhou that served palm civets and had the live animals in the entrance of the restaurant. These animals included palm civets that tested positive for SARS-like viruses. This small outbreak did not grow further. These findings suggest that live-animal markets could increase the transmission of zoonotic viruses (viruses transmitted from animals to humans) through some of the species sold there.

In 2004, further studies indicated that SARS-CoV-like viruses were circulating in various species of bats in different regions of China. Genomic characterization of these viruses indicated a close similarity to SARS-CoV and showed a large genetic diversity. Some bat species were widely found to have antibodies to SARS-like viruses, suggesting that bats could be the wildlife reservoir host for these viruses. In particular, horseshoe bats (*Rhinolophus*) were found to have large numbers of SARS-like coronaviruses. Further studies, published in 2017, found that viruses sampled from bats living in caves in Yunnan, in southwestern China, contained the different pieces of genomic material necessary for generating a SARS-CoV virus through a recombination process.

These findings suggest a possible scenario in which bat viruses can recombine and mutate to generate viruses able to infect humans (Figure 6.2). This could

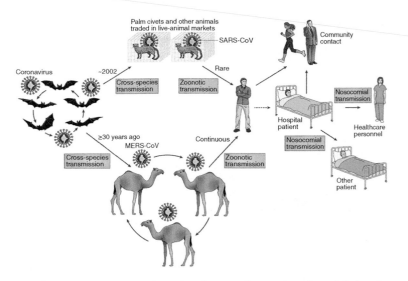

Figure 6.2 Flow of betacoronaviruses from bats to other species to humans before the COVID-19 pandemic. Bats are thought to be the main reservoir of coronaviruses. Sampling of viruses from different bat species has shown a large diversity of coronaviruses, with frequent recombinations resulting in a mixing pot of coronaviruses. These viruses include the related ancestors to the viruses that caused the outbreak of SARS in 2002, MERS in 2012, and the COVID-19 in 2019. The SARS-like bat viruses are thought to be able to infect humans through an intermediate species, potentially in live-animal markets in China, as some of those animals, like masked palm civets, have been found to be infected with SARS-like viruses. The viruses that caused the MERS outbreak are suspected to have passed from bats to dromedary camels several decades ago, leading to the circulation of MERS-CoV in dromedary camels. Several events of MERS-CoV transmission from camels to humans have been reported. Once in humans, both SARS-CoV and MERS-CoV are able to infect other people, mostly in hospitals. Whereas the virus causing COVID-19 probably followed a path similar to SARS-CoV – from bats to humans, and probably through an intermediate host – once the virus was in humans it was able to expand quickly beyond healthcare settings.

have happened through a mediating species, such as the palm civet. However, this is not necessary, as some bat SARS-CoV-like viruses are able to infect human cells, suggesting that a direct infection of SARS-CoV-like viruses is possible without any other mediating species.

It is interesting to note that, unlike MERS-CoV, where several introductions from camels have occurred, SARS-CoV and SARS-CoV-2 both seem to have originated from a single introduction into humans.

Were There Any Super-Spreaders?

Super-spreaders are infected individuals who are able to infect a large number of contacts. It is interesting to point out that a single summary statistic, such as the basic reproductive number (R_0), does not capture the notion of super-spreaders. For instance, a disease with an R_0 of 1 could be due to individuals who are able to infect just one other individual, or to a single individual spreading to many and most others spreading to none. Super-spreaders have been identified as deviations in the number of infections from the expected number. By estimating the R_0 of an infectious disease, one can calculate the maximum number of infected people from an infected individual. Deviations from that number indicate the presence of super-spreading. For instance, if the R_0 is estimated to be 1, the probability that a single individual infects more than 10 would be very low, less than one in a million. If we find a few individuals infecting more than 10 cases in a disease with an R_0 of 1, it would suggest that there are super-spreading events.

The notion of super-spreaders in SARS came after a few cases that managed to infect a large number of individuals, like the doctor who visited Hong Kong from Guangdong in February 2003. The doctor was treating SARS patients in a main hospital in Guangzhou, the capital of the province of Guangdong. Sixteen other guests in the same hotel in Hong Kong got infected. The doctor was taken to a major hospital in Hong Kong, where he infected more than 100 individuals. He died on March 4 after developing symptoms at the end of February. When the SARS cases arrived in Singapore, five more cases of super-spreaders were reported, infecting more than 10 contacts, including family members and hospital workers and visitors.

Super-spreaders have been reported in several outbreaks, including cases of MERS-CoV in 2015 in South Korea, where a single individual coming from the Middle East infected 29 people, mostly in the hospital where he was admitted. Two of these infected individuals were able to infect another 106 people, suggesting that the chain of infection was maintained by a few infected individuals.

It is unclear what the main factors are that could give rise to super-spreading. It could be that the viruses infecting super-spreaders acquire some mutations that make them more prone to infection. There is little evidence for this. Other factors could be the genetic or the clinical history of the super-spreaders, who could have genetic variants or a weakened immune system that makes them prone to producing more viruses. Other more likely factors are environmental, including close proximity with many contacts in mass transportation, housing complexes, and hospitals, and air recirculation in closed settings.

What Are the Clinical Characteristics of SARS and How Does It Compare to COVID-19?

The incubation period of SARS was estimated to be around 5 days, varying between 2 and 10 days, which is similar to COVID-19. The main symptoms include fever and muscle pain, with rare upper respiratory tract infections (nasal congestion or sore throat). Most infections resulted in serious lower respiratory tract infections, but intestinal infections and diarrhea were also reported. A significant number of infected individuals required hospitalization, including up to 20% who required intensive care. In the very early phases of the SARS epidemic it was found that the case fatality rate was 2%. Later, when all cases resolved and it was found that there were not many mild cases, the case fatality rate estimates increased to 10%. COVID-19, in contrast, seems a milder disease, with more mild and asymptomatic cases and transmission, lower case fatality rates (2%), and fewer severe cases and cases requiring hospitalization and intensive care.

Asymptomatic or mild infections were not very common in SARS. Serological studies, looking for antibodies indicating previous exposures to the virus in health workers, revealed very few exposed personnel who did not develop clinical symptoms. The number of unreported cases of COVID-19 is estimated to be high (80%), which was not the case for SARS.

The pathology of SARS is not fully understood. Several factors have been invoked, including the lower respiratory tract infection and an abnormal immune response. It is interesting to note that in some cases of SARS the symptoms became more severe as the virus cleared, suggesting that lung injuries could have been produced through immune mechanisms in the process of virus clearance. Strong upregulation of pro-inflammatory molecules, indicating a hyperactive immune system, was associated with worse progno-sis. In addition, abnormally low levels of white blood cells, involved in immune response, were observed in many SARS patients. Some of these effects have been observed in severe COVID-19 cases, suggesting a common immune deregulation mechanism that leads to complications in both diseases. In SARS it was speculated that the binding of the virus to the ACE2 receptor could have affected its normal activity and could have contributed to the lung injuries observed in the disease.

More men than women were diagnosed with severe disease, and severity and death increased dramatically with age. Different explanations could include lifestyle factors, environmental exposure, or biological differences. The case fatality rate in the elderly – those over 60 – went up to 50%. Children also did not show severe disease. Many of these observations were replicated in animal studies. Mouse experiments show that young mice were less likely to develop disease, and older mice, mostly male, succumbed to the disease. Reduction of female hormone levels in mice led to worse disease, suggesting that hormonal regulation could play a role in the severity of the disease. The parallels in the morbidity and mortality between different species suggests that the age and sex distribution observed in SARS and COVID-19 may have a biological origin.

How Different Are SARS-CoV and SARS-CoV-2 and Their Diseases?

The resemblance between SARS-CoV and SARS-CoV-2 is striking. Genetically they are very similar, with similar genes and using the same port of entry (the ACE2 receptor) to the host cells. The International Committee on Taxonomy of Viruses (ICTV) determined that these two viruses belong to the same viral species, the *severe acute respiratory syndrome-related coronavirus*. This species also contains viruses infecting other hosts, such as SARS-like corona-viruses founds in civets and bats. The reference to SARS in the name of the species reflects strains that are genetically related to the prototypic virus in the species: SARS-CoV, the causative agent of SARS.

The diseases SARS and COVID-19 also share many characteristics. They have very similar incubation time (five days), initial symptoms (fever and respiratory difficulty), and further complications in the lower respiratory tract leading to pneumonia. The parallels in age and sex distributions observed in the two diseases are also remarkable. There are, however, differences in some clinical recommendations, the protocols of testing, the infection fatality rate, and the ease of transmission that probably led to the different success in containment between the two viruses. Some of these factors contributed to the decision by the WHO to choose two different names for the diseases: SARS, the disease caused by SARS-CoV, and COVID-19, the disease caused by SARS-CoV-2.

How Did SARS Disappear?

After initial propagation from the first identified cases in Guangdong in November 2002, the SARS outbreak infected more than 8000 people, resulting in 774 deaths. The rapid spread across the globe to close to 20 countries mostly occurred in outbreaks within hospitals and clusters. The rapid identification and efficient containment measures led to efficient control, and on July 9, 2003, the WHO declared the disease contained. A few isolated cases and laboratory accidents since then have occurred without leading to any significant outbreak.

Many people who recovered from SARS suffered long-term after effects, months and even years after, including lower health status and exercise capacity, in addition to behavioral consequences of stress.

How Was SARS-CoV-2 Able to Propagate While SARS-CoV Was Controlled?

Early detection and isolation worked for SARS-CoV, but has proven to be insufficient to contain the spread of SARS-CoV-2. They are thought to propagate in a similar way, through small droplets of water and infected surfaces, with similar lifetimes in the environment. In the SARS outbreak it seems that most cases occurred within close proximity and in healthcare facilities, but COVID-19 seems to spread more easily. There are, however, several potential differences. One is that transmission of SARS-CoV occurred for the most part only several days after the disease was apparent, whereas SARS-CoV-2 shedding occurs soon after symptom onset, and in a similar fashion whether the case is mildly symptomatic or severe. People infected by SARS-CoV-2 may not be aware of it because symptoms have not yet developed yet or because they are mild. They could, however, still infect other people and contribute to spread of the disease. Other differences could be associated with the differential binding affinity to the receptor of the cell, which could, for instance, contribute to the different patterns of transmissibility between the two viruses.

In the next chapter I will be talking about another virus that has caused several pandemics: influenza. The influenza virus, more familiar to all of us, circulates around the world in a seasonal fashion, but every 30 years or so a new variant appears that is able to rapidly infect the world. The Spanish Influenza of 1918 has been used repeatedly as a comparison to the current pandemic. The last influenza pandemic was first reported in March 2009, and with the same rapidity as the 2019 coronavirus spread around the world in a few months.

7 How Does the COVID-19 Outbreak Compare to Seasonal and Pandemic Influenza?

COVID-19 is influenza on steroids!

Steven Magee

In the beginning of the COVID-19 pandemic, when people were trying to understand the severity of the disease, many comparisons were drawn between this disease and influenza. These comparisons have been a major cause of confusion and misinterpretation. Comparisons with seasonal flu, the influenza virus that comes every winter, led to the idea that the severity of the disease was similar, not taking into account that the virus that causes COVID-19 is new in the population, or that, unlike influenza, no vaccine or efficient antiviral treatment is known. The other comparison was with pandemic influenza, in particular the Spanish Influenza that caused tens of millions of deaths in 1918. The virus responsible for the 1918 pandemic was new in the population, expanded quickly, and caused a significant number of deaths in young adults. That was a time of global war, when influenza viruses were not even known to be the causative agent, and treatments were less developed.

In this chapter, I explain whether influenza can be compared to coronaviruses. I discuss pandemic and seasonal influenza, and how the two diseases compare. I also discuss some of the historical examples of how, in 1918, different systems dealt with the surge of cases and the overloading of the healthcare system.

Is the Influenza Virus Related to Coronavirus?

No, definitely not. These are very different viruses. Influenza viruses are similar in size to coronaviruses, but the genome of influenza is much smaller, with only 13,000 nucleotides split into eight different segments (Figure 7.1).

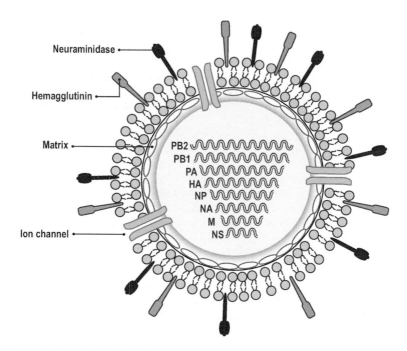

Figure 7.1   **Structure of the influenza A virus and how viruses can be reassorted to generate new genetic variants.** Influenza A is a segmented RNA virus, with its small genome of 13,000 bases split into eight different segments. Each segment codes for one or two genes. The three longest segments code for the replication machinery, the polymerase, composed of three different proteins: PB2, PB1, and PA. The genome is covered by a nucleoprotein that protects the RNA. The genome is inside a capsid made up of the matrix protein. Like coronaviruses, influenza viruses are enveloped viruses and are covered by a lipid bilayer membrane taken from the cells they infect. In the surface of the virus there are two proteins, the hemagglutinin and neuraminidase. These proteins are usually used to give the name to the virus. For instance, a H1N1 virus has a H1 type of hemagglutinin and an N1 type of neuraminidase.

The replication strategy, the way of entering the human cells, and the range of hosts are very different from SARS-CoV-2.

What Is Pandemic Influenza?

The natural hosts of influenza A are aquatic birds, including ducks, swans, geese, and gulls (Figure 7.2). These birds are usually found infected with influenza A viruses, but they are asymptomatic. However, influenza A is also found in mammals including pigs, horses, seals, and humans. Generally, influenza viruses infecting a host are not able to infect other hosts. A pandemic influenza is a virus residing in a non-human host that acquires the ability to infect humans and adapt to humans, expanding in the human population across the world. How do these viruses acquire the ability to infect and

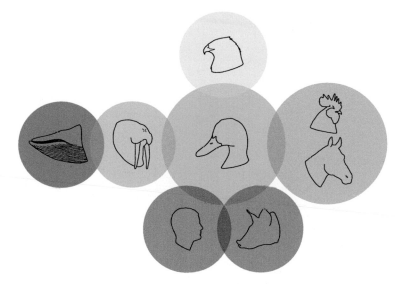

Figure 7.2 Influenza A viruses can be found in many hosts. The most common hosts are aquatic birds. Some aquatic birds, such as ducks, geese, and swans, are frequently infected with influenza A viruses. These infections are usually asymptomatic and do not cause disease. However, a few of these viruses can mutate into highly pathogenic forms that can be deadly for these animals. For instance, H5N1 viruses can kill birds and humans. Influenza A viruses can also be found in swine, horses, bats, and humans, among many others.

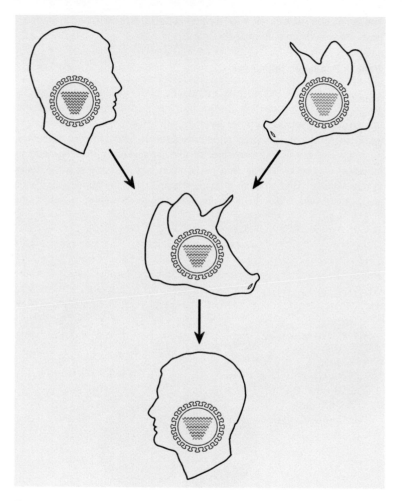

Figure 7.3 Reassortments generate new influenza genetic variants. Influenza A viruses can generate new variants by combining the genomic information from several different strains in a process called reassortment. Reassortment occurs when two different viruses infect the same cell and generate progeny containing segments from the different parental strains. The reassortment process can give rise to new viruses in a host by incorporating segments from viruses circulating in a different host. In the example given here, a mix of a human and a swine virus co-infects the same cell in a pig, generating a new virus able to infect humans. This is the main evolutionary mechanism that gives rise to influenza pandemic viruses.

propagate in humans? Influenza viruses evolve through two main mechanisms. The first is the accumulation of mutations. The second is a process called reassortment, in which two viruses are able to generate progeny containing genomic material from both of them. Reassortments in influenza are like recombinations in coronaviruses – a way of combining genetic material from different viruses. Reassortment is a particular way of generating new variants in viruses with segmented genomes – genomes made up of different independent fragments or segments. While coronaviruses have the genome in a single strand, segmented viruses, such as influenza, have the genome split into fragments. New viruses can be generated by different combinations of these fragments. Through reassortment, new influenza viruses are able to acquire genes from viruses that are able to infect humans (Figure 7.3).

There have been three influenza pandemics in the twentieth century (in 1918, 1959, and 1968), and another one in 2009 (Figure 7.4). The pandemic of 1918, the Spanish Influenza,[1] caused 50–100 million deaths around the world. Given that the population of the world at that time was 1.8 billion people, the total population fatality is estimated to be close to 3%. By contrast with SARS-CoV-2, it mainly affected a younger population, people in their twenties to forties. At the time the situation was captured in pictures (Figure 7.5) and narrated in books and letters, such as this one sent from a physician working in a camp near Boston:

> This epidemic started about four weeks ago, and has developed so rapidly that the camp is demoralized and all ordinary work is held up till it has passed . . . These men start with what appears to be an attack of la grippe or influenza, and when brought to the hospital they very rapidly develop the most vicious type of pneumonia that has ever been seen. Two hours after admission they have the mahogany spots over

---

[1] The influenza pandemic of 1918 was popularly called the "Spanish Flu" as Spain was the first country to report in the free press the effects of the pandemic, when even the king became infected. Many prior cases occurred in several countries fighting in the First World War, but were not reported due to war-imposed censorship. Nowadays names of infectious diseases are chosen with more political finesse by international organizations.

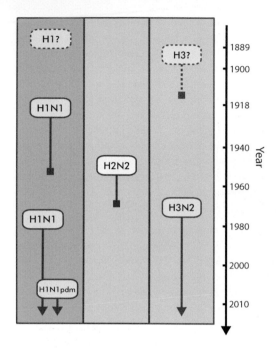

Figure 7.4   Influenza pandemics in the twentieth and twenty-first centuries.
Approximately every 30 years a new influenza virus appears in the population and
causes a pandemic. The first well-reported case was the H1N1 1918 Spanish Influenza.
The origin of the virus is still unclear. The virus came in several waves: spring 1918, fall
1918, and spring 1919. The effects of this pandemic were catastrophic, with 50 million
deaths worldwide. After that, the virus circulated in the population in a seasonal fashion
with a much lower infection fatality rate. The virus circulated in this fashion until 1959,
when a new pandemic strain, the H2N2 Asian Influenza, replaced the H1N1 virus. The
H2N2 pandemic was relatively milder, but still reached more than one million deaths. The
H2N2 virus was a reassortant from the H1N1 circulating virus and a virus of avian
origin. The H2N2 virus then circulated seasonally until 1968, when a new pandemic
strain, the H3N2 Hong Kong Influenza, created a new pandemic, with a similar number
of fatalities as the previous one. This H3N2 virus was also a reassortant, a mixing of an
H2N2 virus circulating in humans with a virus of avian origin. The H3N2 become
seasonal, replacing all other strains. In 1977, an H1N1 virus, similar to the H1N1 viruses
circulating in the 1950s, reappeared in the population. Since the end of the 1970s, two
different influenza A strains have been circulating in the population: H3N2 and H1N1.
In 2009, a new pandemic strain of swine origin rapidly spread around the world.

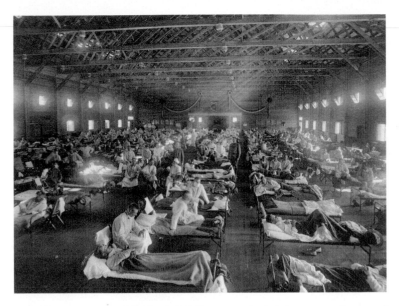

Figure 7.5   **The rapid spread of the 1918 influenza pandemic led to a surge of cases and deaths that overwhelmed the healthcare system.** The number of sick people that needed attention saturated hospitals and funeral homes. Improvised field hospitals and care centers were deployed around the world. In addition to the overloading of hospitals, many US healthcare workers were deployed in Europe assisting the troops fighting in World War I. Many volunteers put their lives at risk to help take care of the sick and bury the dead. This photograph was taken in an emergency hospital during the influenza epidemic in Camp Funston, Kansas.

the cheek bones, and a few hours later you can begin to see the cyanosis extending from their ears and spreading all over the face, until it is hard to distinguish the coloured men from the white. It is only a matter of a few hours then until death comes, and it is simply a struggle for air until they suffocate. It is horrible ... We have been averaging about 100 deaths per day, and still keeping it up ... We have lost an outrageous number of nurses and doctors ... It takes special trains to carry away the dead. For several days there were no coffins and the bodies piled up some-thing fierce, we used to go down to the morgue (which is just back of my ward) and look at the boys laid out in long rows. It beats any sight they ever had in France after a battle. Good-by old Pal, God be with you till we meet again.

Just after the First World War, this pandemic led to dramatic public health measures similar to those SARS-CoV-2 is prompting now, such as the prevention of mass gatherings. But the infection fatality rates of the 1918 pandemic are estimated to be near 10%, quite a bit higher than currently reported for the SARS-CoV-2 case fatality rates and probably much higher than the infection fatality rates. The influenza pandemic came in several waves. A first wave in spring 1918 presented a milder disease similar to seasonal influenza. The deadliest wave started in August 1918 and quickly expanded around the world, lasting for two months. In New York, it was declared over on November 19, 1918. The number of deaths in New York City alone was estimated to be around 30,000, and in the whole USA about 675,000. A third, milder wave came in the winter of 1918 and spring of 1919, with 67 deaths reported in New York City.

The pandemic in 1957 was a result of a reassortment of the seasonal descendants of the 1918 pandemic with a virus related to influenza A viruses circulating in birds. The number of deaths was estimated to be more than one million worldwide. Similar numbers were reported in the pandemic of 1968, resulting from another mix of human and avian-origin viruses.

The most recent influenza pandemic occurred in 2009, and had its origin in another viral mixing between viruses infecting pigs. Despite fewer deaths than previous pandemics (approximately 200,000 worldwide), 80% of cases were reported to be in the younger population.

What Is Seasonal Influenza?

Once a pandemic virus has entered the human population, the virus becomes seasonal, coming back every winter with slightly different characteristics (Figure 7.6). Influenza viruses accumulate mutations at a high rate, roughly 20 mutations per year. Immune responses from one year could be suboptimal or ineffective in the next. This continuing evolution poses a challenge for vaccine developers; strains selected for the vaccine need to be updated on a yearly basis.

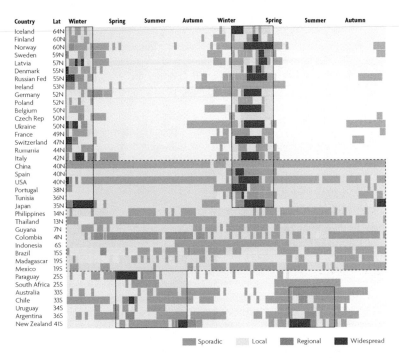

Figure 7.6 **Seasonal influenza follows a yearly periodic pattern of infections with a higher incidence in winter months.** The figure shows the number of influenza-like illnesses (ILIs) from the World Health Organization's surveillance system in different seasons. The countries are organized by latitude, with northern countries at the top and southern in the lower part. In northern Europe, all cases of ILI can be found in winter or early spring, mostly between December and March. In summer there are no reported cases. In the southern hemisphere the peak of cases is observed in June to September, corresponding to the winter months. In other countries located between the tropics, the cases of influenza are distributed more homogeneously, with weaker peaks and not following a very strong seasonal pattern. The cases of influenza around the globe peak similarly in countries of the same latitude.

The numbers of hospitalizations and deaths from seasonal influenza vary year to year. The Centers for Disease Control and Prevention estimated that in the 2018–2019 season in the USA, 35.5 million became sick with seasonal influenza viruses, leading to almost half a million hospitalizations and 34,200 deaths.

How Does Influenza Transmit and Cause Disease?

Influenza has two main means of transmission: through the air in droplets of water released from sneezing or coughing, and on surfaces. How effective the transmission is depends on atmospheric conditions, including temperature and humidity. These factors are thought to play a role in the seasonal effects observed in seasonal influenza.

Influenza infections mostly happen in the upper respiratory tract, leading to a variety of symptoms, including fever, cough, sore throat, muscle pain, and fatigue. In some cases, the infection can lead to complications, including death. It has been estimated that around half a million people in the world die from seasonal influenza every year.

How Does Influenza Virus and Illness Compare with SARS-CoV-2 and COVID-19?

One of the most unfortunate and confusing metaphors from the beginning of the COVID-19 outbreak has been the comparison between the new pathogen and influenza, both seasonal and pandemic. Although the comparison can be illustrative, it can also be misleading.

Influenza and coronaviruses are two very different viruses, belonging to two very different families and having very different means of entering cells and replicating. They also encode their genomic material in different ways, and the proteins and genes of the two viruses have no resemblance to each other. They have different incubation periods – a couple of days for influenza and five or more for SARS-CoV-2 – and the pattern of infections is different. Seasonal influenza is mostly a disease of the upper respiratory tract. Although complications and pneumonias can occur, they are not as common as in COVID-19. Seasonal influenza morbidity and mortality are associated with the very young and the old, whereas COVID-19 cases in the young population are rare. There are specific drugs and vaccines for seasonal influenza, whereas no drugs or vaccines are available for the 2019–2020 COVID-19 outbreak.

Perhaps a better comparison could be drawn between the influenza pandemic and the current COVID-19 pandemic, as there was then, and is now, no vaccine available, although some drugs may be useful. The closest comparisons may be the influenza pandemics of 1918, 1957, and 1967.

However, the comparison with the 1918 pandemic could also be misleading, given that the world was coming out of a devastating war, information was partly suppressed under the war-time conditions, and the health system was not as developed as it is today. In 1918, scientific knowledge was very limited (it was not known that the cause of the pandemic was a virus, and most doctors believed it was a bacterial disease), and the capacity to develop vaccines and drugs was non-existent.

There are, of course, some similarities between influenza and SARS-CoV-2, by virtue of transmissibility of infection through droplets of water and surfaces. One important and useful comparison with pandemic influenza is that the population had not seen the virus before, and it can create pandemics. The high number of susceptible people can create a surge in the number of cases, overwhelming the healthcare system. It is interesting that mitigation measures of isolation, quarantine, closing of schools and public places, and social distancing are the same a century later.

What Did We Learn from the Public Health Measures Taken in the 1918 Pandemic Flu?

Although pandemic influenza and SARS-CoV-2 are very different viruses, the containment measures adopted in 1918 to retard the spread of the Spanish Influenza could illustrate how different public health measures can mitigate or exacerbate the mortality of the disease.

In mid-September 1918, the Spanish Influenza was spreading quickly through the military camps. To support the war effort in Europe, Philadelphia decided to organize a big parade on September 28. A few days later, all hospitals in Philadelphia were full and thousands were dead within a week. In St. Louis, the response was quite different. Even before the first cases were identified, the city banned public gatherings and closed schools. That significantly reduced both the surge in burden on hospitals and the mortality rate. One of the most interesting lessons from the study of the impact of public health measures during the influenza pandemic of 1918 is that public health interventions have an important influence on mortality and the ability of the healthcare system to take care of the surge in cases.

The 1918 pandemic led many governments in Europe to adopt different forms of socialized, centralized healthcare systems. Governments felt compelled to create special offices and ministries, and to increase funding for public health. It was also clear that large pandemics were unaffected by local or national boundaries, and that efforts needed to be coordinated at an international level. We have to remember that 1918 was the year that the First World War ended, and there was a political will to avoid similar international disasters.

In 1920 the League of Nations, the precursor of the United Nations, created the Health Organization of the League of Nations. In 1946, after the Second World War, it became a United Nations specialized agency, the World Health Organization.

8 Are There Therapeutic Options?

Innovation comes only from an assault on the unknown.

Sydney Brenner

All who drink of this remedy recover in a short time, except those whom it does not help, who all die. Therefore it is obvious that it fails only in incurable cases.

Galen

At the time of writing there are no vaccines or specific antiviral therapies for the SARS-CoV-2 virus with significant reductions in mortality. Lessons learned from the SARS outbreak have definitely helped us to understand many aspects of this emerging virus and the associated pathologies. But no effective therapies have been developed for SARS that could be leveraged for the COVID-19 outbreaks.

In this chapter, I briefly discuss some of the ideas that have been suggested and are currently being tested. I start with a simple primer on how the virus is diagnosed and how infections are currently treated depending on the severity of the disease. I also discuss some of the therapeutic strategies that are currently being tried, including vaccines.

How Is the Virus Diagnosed?

Testing is a highly efficient method to allocate healthcare resources to people who will be in need and to take measures to reduce the chance of infecting others. The virus was isolated from a sample from Wuhan on January 7, 2020, and the genomic sequences were shared with the international community on January 12. That allowed the rapid design of testing tools based on polymerase

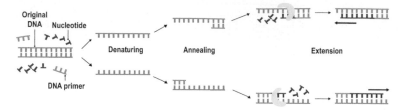

Figure 8.1 Polymerase chain reaction or PCR is a technique for exponentially amplifying the amount of genomic material. The process starts with two primers, design to complement the genomic region of interest. For instance, if we are interested in a particular gene, we will design the primers to be specific to the sequence of that gene. The sample DNA is heated so the two strands of DNA are separated. The primers will bind to the DNA of interest. A polymerase will extend the sections of the genome that bind to the specific primers, creating two copies per copy of DNA of the gene present. If the process is repeated through several cycles, the amount of DNA will amplify exponentially. Small amounts of genomic material can then rapidly expanded to large amounts.

chain reaction, or PCR (Figure 8.1). This is a procedure for amplifying genomic material, in which specific pieces of genomic material are copied, the copies are copied again, and so on for several cycles, until the amount is large enough for identification. PCR tests measure directly the presence of the genomic material of the virus, but not the immune response to it. PCR tests require laboratory equipment and can take a few hours to generate results.

A couple of weeks after a person has been infected with SARS-CoV-2, the immune system develops antibodies – molecules that are able to recognize the virus very specifically. The identification of specific antibodies to the virus indicates that a person has been infected in the past and has been able to mount an effective immune response. Serology tests examine the blood for the presence of these antibodies. Because the generation of antibodies in an infected person takes a couple of weeks, antibody-based tests cannot provide a rapid assessment of the presence of an active virus. However, these tests are very useful to determine whether a person has been infected in the past, even if the virus is no longer there.

Serology tests are very useful for assessing how many people have been infected but not reported as such, due mostly to the mild or unobservable effects of the infection (Figure 8.2). Most of our estimates in the beginning of

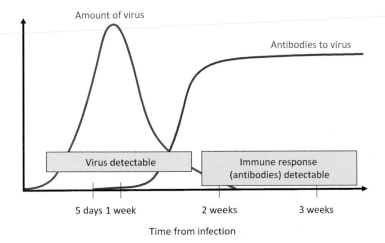

Figure 8.2   **Graph representing the levels of virus and antibodies from a patient with a mild infection.** Viruses can be detected a few days after infection. The incubation period, when the symptoms of the disease are manifest, is around five days. Both the virus genome and proteins could be detected in the first two weeks after infection, with high variability across infected individuals. The antibodies against the virus are generated around two weeks after infection, and they can be detected in blood. They do not mark whether a person is currently infected with the virus, but whether an immune response has been mounted.

the pandemic were through mathematical modeling, with assumptions and large variance in estimates. Serology surveys could provide a more accurate assessment of true infection rates. They can also indicate that a person has been able to mount an effective response to the virus, and help determine whether the person can actively interact with more at-risk patients. Finally, they can help to estimate the rates of re-infection and the duration of the immune response. The immune responses to some viruses, such as measles, last for a lifetime, whereas the responses to others, including the common coronavirus circulating in humans, last only a couple of years. We do not know yet how long the immune response for SARS-CoV-2 will last.

Immunoassay tests look for viral proteins using lab-generated antibodies that bind specifically to proteins on the surface of the virus. They detect viral proteins directly, but not the response of the immune system, so a positive result indicates the presence of a virus, indicating an active infection. These tests and PCR detect the virus using the genome or viral proteins, but do not say anything about previous infections.

The advantage of the antibody-based tests is that they do not require specialized equipment, such as PCR machines, and they could be deployed easily, providing results in less than an hour.

How Are the Infections Treated?

Current management of COVID-19 is mostly supportive. The WHO provides some guidelines for treatment based on the severity of the disease:

- Mild cases should be monitored, and symptoms (fever, muscle pain) treated with over-the-counter drugs.
- Severe cases require oxygen therapy and treatment of co-infections. Co-infections may require treatment with antibiotics, if bacterial infections are suspected, or with neuraminidase inhibitor if influenza is also suspected.
- Cases presenting sepsis and septic shock: Sepsis is caused by a dysregulated immune response that can lead to organ dysfunction. Septic shock is a severe condition derived from sepsis associated with very low blood pressure and widespread infection. Many of the deaths associated with COVID-19 were related to sepsis.
- Cases presenting acute respiratory distress syndrome (ARDS): ARDS is caused by rapid inflammation in the lungs, causing the accumulation of fluids in the lungs that can preclude normal breathing, with reduction of oxygen reaching different organs. Patients could require ventilators or intubation.

Blood from recovered patients contains antibodies that can neutralize the virus. These antibodies are generated by the adaptive immune system after two weeks, and are highly specific to the virus. The use of plasma from recovered patients to treat actively infected patients has been approved in several countries.

Are There Any Drugs Used to Treat Coronavirus Infections?

Currently there are no specific COVID-19 antiviral drugs approved, but there are several in clinical trials. The first therapeutic strategies in cases requiring immediate action are the repurposing of existing drugs to reduce the severity of the disease. These drugs target different aspects of the virus life cycle, its replication, its interactions with humans, or are designed to help the immune system fight the virus.

A good place to start looking are the therapies that were tried during the SARS outbreaks. Extensive work has been done in animal models, and a variety of different therapeutic strategies in the SARS context have been proposed. Given the genetic similarity between SARS-CoV and SARS-CoV-2, it is likely that we could leverage some of the knowledge and models developed for SARS to accelerate the research into COVID-19 therapies. There is, however, no solid clinical data on the efficacy of potential treatments during that outbreak.

Remdesivir is a broad-spectrum antiviral drug developed as a treatment for the Ebola virus infection, with potential use in other viruses, including corona-viruses. It acts by interfering with the viral replication machinery. It has been shown to be effective *in vitro* and in animals infected with SARS. There are several ongoing clinical trials evaluating its efficacy in humans.

Favilavir is another broad-spectrum antiviral drug approved for treating influenza, and has also now been approved in China for the treatment of COVID-19. It works by targeting the replication machinery of the virus. It has been tested in animal models for many RNA viruses, including influenza, West Nile virus, and Ebola virus, among others.

Drugs used in HIV treatment (lopinavir and ritonavir) are also being studied. *In vitro* experiments and small-scale studies during the SARS outbreak in 2003 indicated a potential reduction in severe outcome (acute respiratory syndrome or death). Although case reports and small studies have indicated a potential benefit for COVID-19 patients, a randomized trial in 199 hospitalized patients in Wuhan did not reveal benefits beyond the standard of care.

Chloroquine and the related hydroxychloroquine, widely used to treat malaria, have been seen to inhibit SARS-CoV and SARS-CoV-2 in cells grown in lab conditions. Besides interfering with viral infection, these compounds can modulate immune responses, that are often dysregulated in severe COVID-19 cases. The effect and mechanism of action in COVID-19 cases is still unclear.

There are many other strategies that are being tested, such as the use of ACE2 decoys that will bind to the virus, competing with the cell ACE2, and strategies to modulate the immune response to the virus to prevent an overacting immune system, including repurposing immune modulators that have been approved to treat immune disorders.

Are There Vaccines for Coronaviruses?

Vaccines work by activating an immune response and have been the most successful way to combat viruses. Vaccines prime the immune system by presenting parts of, or a weakened version of, a pathogen. This stimulates the adaptive immune system to recognize the pathogen. Extensive vaccination can lead to widespread immunity in the population (herd immunity), making it harder for the pathogen to infect individuals and cause a large number of diseases. Vaccines against viruses have been proven to be extremely effective, such as the vaccines for measles, mumps, rubella, smallpox, and hepatitis B, among many others. Systematic vaccination and active international efforts have ended smallpox, one of the most devastating infectious diseases in the history of humanity, which has killed nearly 300 million people in the twentieth century. The efficacy of some vaccines could be variable, like the influenza vaccines, due to the variations in the circulating viral strains.

Currently there are no vaccines for SARS-CoV-2 or SARS-CoV. In fact, there are no vaccines for any human coronaviruses, although there are now clinical trials underway. There are some vaccines for coronaviruses infecting domestic animals, including avian and dog coronaviruses. However, even these vaccines can present a challenge due to the continued emergence of novel serotypes.

There are many challenges ahead. It is important to understand whether any vaccine will help to develop immunity against the virus and how long it will last. Disease enhancement is a phenomenon in which the vaccine could lead to severe immune reactions that can worsen the disease. A study of a SARS vaccine in 2004, in ferrets, showed that the immunized animals developed a strong neutralizing antibody immune response when challenged with the virus, but they also presented inflammatory damage in the liver, leading to hepatitis. Finally, an important challenge is that, as in influenza, the high evolutionary rate of coronaviruses and their ability to recombine generates a continuous source of antigenic variability and a formidable challenge to vaccine development. A vaccine that effectively deals with the circulating diversity of the virus could be challenging, as it is for some of the coronavirus vaccines in animals.

Conclusions

Life can only be understood backwards; but it must be lived forwards.

Søren Kierkegaard

There are many, many important questions that remain to be answered regarding the SARS-CoV-2 viruses:

- Where is this virus coming from and how did it manage to infect humans? Understanding this route could help prevent further events like SARS or COVID-19 outbreaks in the future. Also, it will help to understand the specific mechanisms of adaptation of zoonotic viruses.
- What are the specific mechanisms by which SARS-CoV-2 causes disease? In an important fraction of infected individuals, SARS-CoV-2 causes severe disease, while in others only mild symptoms. Understanding the mechanisms of disease will help us develop and apply therapies to control the severity of the disease.
- How does this virus interact with the immune system? We need to understand the specific mechanisms of normal immune response to the virus, both the innate and adaptive responses. In many patients the virus triggers an abnormal immune response that becomes difficult to control and can kill them. We do not currently understand in detail how the virus interacts with the immune system in mild or severe cases.
- What is the true number of infected people? There have been varying estimates on how many people are really infected with the virus. Modeling suggests that there could be a significant number of infected individuals who do not develop any symptoms, or only develop very mild symptoms. Understanding these numbers could help us to better

understand the severity of the disease, the host factors that can control the progression to disease, and the transmission routes through unreported cases, among many other factors.

- What is the role of mild symptomatic and asymptomatic cases in the spread of the infection? Epidemiological models have suggested that there is a significant number of unreported cases that contribute to the spread of the disease.

- Very low numbers of cases have been reported in children and very low numbers of deaths. Why do most children exhibit only mild symptoms, while most fatalities occur in the elderly and mostly in males?

- Are there genetic or epidemiological markers of severity? Beyond age and other related diseases, there are other indicators to assess whether a person will suffer from serious disease. As in many other infectious diseases, there are genetic components that can contribute to the risk, predisposing some people to be protected or to develop serious disease. We do not yet know if there are genetic markers associated with COVID-19.

- How long will the immune response last after a person has recovered? Is it possible to be re-infected? It seems that after the infection, the generated antibodies are able to neutralize the virus. But in several coronaviruses, the memory of the response lasts only a few months and re-infections with the same virus have been observed. We do not know if that will be the case for COVID-19 or if recovered patients will be able to develop long-lasting immunity. Lack of long-lasting immunity could lead to future waves of COVID-19 unless the virus is eradicated.

- Will any therapy – antiviral drug or vaccine – be able to treat the disease effectively? In the next months and years, we will see the approval of antiviral drugs to treat COVID-19. It is to be hoped that a vaccine will become available too.

- Will society be able to generate herd immunity? If enough individuals develop a lasting immune response, we could make it difficult for the virus to spread again. With current numbers it seems that we need at least half of the population to be infected, or a vaccine, to tamp down the spread of the virus.

- Will there be subsequent waves of the disease? If so, will they be milder or affect different sections of the population? In other pandemics, such as the 1918 influenza, several waves came through the population until the virus settled in as a seasonal influenza.

- Will the virus evolve to escape the therapies or the built immunity? RNA viruses evolve very rapidly. That constitutes a significant challenge to the development of therapy or vaccine. As the virus propagates it will diversify and we will have to make sure that whatever therapies we develop are able to challenge the majority of the circulating viruses.

All the questions above will keep the scientific community very busy, and the next few years will witness many developments. Each gene, each protein, each piece of knowledge about the virus is an opportunity to fight the disease. The scientific community is frenetically searching for ways to block the entrance, the replication, and the release of the virus; to understand the disease; to modulate the immune response; and to evaluate optimal public health measures for this and other many infectious diseases.

We are now in a very different time than when the influenza virus struck in 1918. In 1918 there was neither the technology nor the knowledge that we have today to study the infectious agent rapidly. Many national and international efforts are now in place to answer the questions above. It is likely that some treatments and vaccines now in clinical trials will have some effect on COVID-19 cases. They will not help in the first wave of the disease, but are likely to be available in the near future.

There are many viruses circulating in humans, all of which came at some point from different species. We have to remember that there were four coronaviruses circulating in humans before the SARS-CoV-2 outbreak. Other examples include the influenza viruses starting with the pandemics of 1918, 1957, 1968, and 2009. So, it is probable that as a larger fraction of the population is infected, humans will develop immunity, mitigating the surge in the number of cases and the overloading of healthcare systems. Immunity will also be achieved if efficient vaccination is deployed. It is likely that SARS-CoV-2 is here to stay, like some of its coronavirus cousins.

The emergence of this virus has been only one of many events, although a most disrupting one. The WHO has reported more than 1500 new pathogens since 1970, most of them of animal origin. Emerging viruses in the last 50 years include HIV, the H1N1 influenza pandemic, several Ebola virus outbreaks, the MERS coronavirus, and Zika, among many others.

There are many factors that have contributed to the appearance of recent outbreaks and their expansion in the population. One of the main factors is the growth of the population and the migration to large, overcrowded urban areas. In 1918, when the Spanish Influenza appeared, there were 1.8 billion people in the world; today there are 7.8 billion. The intensity of travel, both domestic and international, has been one of the main factors leading to the rapid spread of this pandemic virus. The outbreak of COVID-19 was reported in Wuhan at the end of December 2019, and the first case in the USA was reported January 20, 2020. By the end of January the virus was reported in 21 countries. A similar situation occurred in 2009, when the first cases of the H1N1 pandemic virus were reported in April, and by the end of May cases were found on all inhabited continents. These recent examples highlight the difficulty of containing emerging respiratory viruses in a modern world. Other factors that have exacerbated the emergence of new pathogens from other species are associated with the severe disruptions of their natural habitats and the contact between domestic livestock and wildlife.

The COVID-19 pandemic has been a very significant and dramatic event in our lives. The dream of globalization that has driven the world for the last 50 years has been shaken by the lockdown of countries, the congealed global economy, and the systematic paralysis of the world flow. It is uncertain what the long-term economic and political consequences could be, but hopefully next time we will be better prepared.

Summary of Common Misunderstandings

The COVID-19 coronaviruses is another type of flu. Coronaviruses and influenza are very different viruses. They do not share genes, proteins, or the same way of infecting cells. The disease that they cause is also different. Most infections of seasonal influenza are in the upper respiratory tract, while COVID-19 causes common lower respiratory tract infections as pneumonias that require hospitalization and intensive care for a fraction of patients.

The COVID-19 mortality rate is like that of seasonal influenza. It is not. The common flu has a varying case fatality rate, but typically it is estimated to be 1 in 1000. COVID-19 estimates are around 20 in 1000. They differ by a factor of 20. It is also spreading more efficiently than seasonal influenza. These two facts make the number of severe cases increase rapidly, overwhelming healthcare facilities and workers. Moreover, there is some protection in the population for seasonal flu, as it is a common circulating virus, but there is none for a new pandemic virus. That means the whole population is at risk. In addition, we have antiviral therapies and vaccines for flu, but nothing of this sort for the pandemic coronavirus.

The COVID-19 mortality is like that of the 1918 influenza. It is still unclear what the total number of deaths associated with COVID-19 will be. The infection fatality rate of the 1918 influenza virus was estimated to be near 10% – much higher than any of the current estimates for COVID-19.

The virus causing COVID-19 is not related to the virus causing SARS, because the former caused a pandemic whereas the latter did not. The viruses causing SARS and COVID-19 are related in many ways. They are both within the same viral species among the betacoronaviruses. They have the same genes and similar proteins. They use the same entry point to the cell, the ACE2 receptor. The diseases also present some similarities, with common lower tract infections and severe respiratory complications. The case fatality rate of COVID-19 is lower, but the infectivity is higher than for SARS. The similarity between the two viruses helps us to use the previous clinical experience, scientific knowledge, and techniques from the SARS 2003 outbreak to accelerate the search for therapeutics.

Children and young adults do not get infected. Children and young adults get infected but, in most cases, they get a milder disease. Occasionally they get severe complications, including death. In addition, infected people, independent of age, contribute to the spread of the disease. There is a moral responsibility to care for all and particularly for the most vulnerable in our society.

Containment measures are ineffective. The highly proactive response taken by Asian countries (China, Hong Kong, Taiwan, Singapore, South Korea) in the first two months of the pandemic have proven that containment measures are effective. Rapid identification and testing of cases, contact tracing, early isolation, and treatment of infected cases together with social distancing measures reduced the number of infected cases.

When the number of cases is too high, strong public health mitigation measures can reduce the spread and the sudden surge of cases requiring medical attention. The surge of medical cases leads to an overwhelmed healthcare system and large numbers of unprotected healthcare workers unable to provide clinical care for patients of any disease. This increases the complications and deaths associated with COVID-19 and any other disease.

Historical evidence from previous pandemics has shown that social distancing and early public health measures can reduce the surge of cases. Examples from during the 1918 influenza have shown that cities that do not effectively implement mitigation strategies can suffer a dramatic increase in the number of cases and deaths.

The coronavirus is of extraterrestrial origin, and other interesting ideas. There are many questionable theories that have populated the Internet with increasing degrees of sophistication and creativeness. For instance, it has been suggested that the virus came to Earth in a meteorite or a comet that fell over northeast China in the fall of 2019. This theory follows the unorthodox line of thought described in a book published by some cosmologists in 1979, arguing that pandemics and viral outbreaks are of extraterrestrial origin. The same hypothesis was suggested by the same people about SARS in 2003 and about COVID-19 in 2020.

This theory is easily refutable, as from the genomic analysis of the virus we know that the virus is very closely related to many other coronaviruses in bats and other mammals. It has followed the natural evolutionary processes – mutations and recombinations – observed in coronaviruses. Natural evolution from an animal reservoir is the most parsimonious explanation. Why would a comet bring to China a virus that is related to other coronaviruses circulating in Chinese bats? Is it not more likely that it is coming from another mammal?

The same reasoning goes for all the viral outbreaks, including SARS, the flu pandemics of 1918, 1957, 1968, and 2009, Ebola, etc. All of those viruses are related to viruses circulating in animals, and animals are the most likely origin.

Updates at Press

Number of Infected

- Serology tests provide a first glimpse of how many people have mounted an immune response detectable by these tests. These numbers could be as high as 25% in heavily hit areas, such as New York City. As serology tests become widely available, we will get better estimates of these numbers and the true number of unreported infections.

Risk of Severity

- The initial observation that COVID-19 is more frequent in the older population and in men has been replicated in many other countries as the number of worldwide infections has increased. However, there have been some differences. For instance, more young patients have been admitted in hospitals in the USA compared to other countries, associated with risk factors that include obesity and diabetes (obesity in the USA reaches 40%, versus 20% in South Europe and 6% in China).

Other Clinical Manifestations: As the Number of Infections Increases, a Variety of Clinical Manifestations Have Been Reported

- It is now clear that COVID-19 affects many organs beyond the lungs, including the heart, digestive system, kidneys, and skin. Blood clots have been frequently seen in severe cases, leading to severe complications including strokes. Skin conditions include unusual rash and chilblains in toes (sometimes named COVID toes), mostly found in the young population.

- A hyperinflammatory disease, similar to Kawasaki disease, has been reported in children, resulting in a small numbers of deaths. Symptoms include high fever, pain, and gastrointestinal symptoms. Preliminary reports indicate that this condition could be more frequent in boys than girls.
- The loss of smell and taste has been widely reported to be one of the common telltale manifestations of SARS-CoV-2 infections. The mechanism, how the virus causes the loss, is unclear.

Therapeutic Developments

- The first results of a large clinical trial of Remdesivir with more than 1000 individuals has shown that it could reduce the time of infection by four days, but not a drastic reduction in COVID-19-associated mortality. Based on these results the US Food and Drug Administration issued an "emergency use authorization" for the treatment of SARS-CoV-2 infections.
- Several drugs tried for SARS-CoV and MERS-CoV have been repurposed to treat SARS-CoV-2 infections. A clinical trial in Hong Kong with combination treatment of interferon beta-1b, lopinavir–ritonavir, and ribavirin resulted in shorter time (five days) of viral shedding than control groups.
- Early observations in small hydroxychloroquine clinical trials have not been supported by larger studies, and potential adverse events have been reported, including heart problems.
- But clinical trials continue and hopefully we will have more positive news soon.

References

www.thelancet.com/journals/lancet/article/PIIS0140-6736(20)31024-2/fulltext

www.thelancet.com/journals/lancet/article/PIIS0140-6736(20)31094-1/fulltext

www.nature.com/articles/d41586-020-01295-8

www.thelancet.com/journals/lancet/article/PIIS0140-6736(20)31042-4/fulltext

www.nejm.org/doi/full/10.1056/NEJMoa2012410?query=featured_home

Suggested Further Reading

Here, I suggest further information for the enthusiastic reader through a curated guide organized by chapters and topics. References include popular books, textbooks in related topics, scientific articles, and scientifically sound blogs.

Chapter 2: How Is the Coronavirus Spreading?

Lu, R., Zhao, X., Li, J., et al. Genomic characterisation and epidemiology of 2019 novel coronavirus: implications for virus origins and receptor binding. *Lancet*, doi: 10.1016/S0140-6736(20)30251-8 (2020).

This is one of the first articles reporting the COVID-19 outbreak in Wuhan, China. The paper narrates the story of the first cases, how the coronavirus was found, the description of its genome, and how it relates to other coronavirus.

Gordis, L. *Epidemiology*, 5th ed. (Elsevier/Saunders, 2014).

This is a classic text book in epidemiology. It provides a theoretical foundation as well as practical cases with real-world examples. It is highly pedagogical and recommended for newcomers.

Novel Coronavirus Pneumonia Emergency Response Epidemiology Team. [The epidemiological characteristics of an outbreak of 2019 novel coronavirus diseases (COVID-19) in China]. *Zhonghua Liu Xing Bing Xue Za Zhi* 41, 145–151, doi:10.3760/cma.j.issn.0254-6450.2020.02.003 (2020).

A very detailed analysis of COVID-19 cases in China from the beginning of the outbreak till mid-February. The analysis includes a description of the first patients, including age and sex, characteristics, estimates of case fatality, and spread rates. A little technical, but with many interesting numbers.

World Health Organization, *Sex, Gender and Influenza* (World Health Organization, 2010).

This is a nice report by the WHO studying the different responses of men and women to influenza infections and disease severity. Genetic and hormonal differences contribute to observed differences in disease and mortality of respiratory viruses.

Chapter 3: What Is a Coronavirus?

Flint, S. J., Racaniello, V. R., Rall, G. F., Skalka, A. M., & Enquist, L. W. *Principles of Virology*, 4th ed. (ASM Press, 2015).

One of my favorite textbooks about viruses. The two-volume *Principles of Virology* organizes the main common characteristics by the general principles of how they infect the cell, replicate, and interact with the immune system. It has good illustrations and clear explanations.

Levine, A. J. *Viruses* (W.H. Freeman and Co., 1992).

Arnold J. Levine wrote this popular book about viruses in 1992. It offers a great introduction to the history, the biology, and the diseases caused by viruses. Beautifully written, concise, and interesting stories make it one of the best introductions to the world of viruses.

Morse, S. S. *Emerging Viruses* (Oxford University Press, 1993).

Steve Morse compiled the expert opinion of scientists in the field of emerging viruses. Written before SARS and COVID-19, it provides a very nice perspective on how viruses emerge into the human population, how they cause disease, and the measures for prevention. The message is the same today.

Knipe, D. M. & Howley, P. M. *Fields Virology*, 6th ed. (Wolters Kluwer/Lippincott Williams & Wilkins Health, 2013).

Fields is *the* classic text in virology for those who want to learn about the specific details of viral families. A good chapter on coronaviruses provides technical details on the molecular biology, disease, and recent outbreak.

Coronaviridae Study Group of the International Committee on Taxonomy of Viruses. The species Severe acute respiratory syndrome-related coronavirus: classifying 2019-nCoV and naming it SARS-CoV-2. *Nature Microbiology*, doi:10.1038/s41564-020-0695-z (2020).

This is the article by the International Committee on Taxonomy of Viruses in which the name SARS-CoV-2 is proposed for the agent of COVID-19. It gives a perspective about the virus, and why it should be located within the same viral species as the SARS viruses.

Chapter 4: How Is the Coronavirus Changing?

Rambaut, A. Phylogenetic analysis of nCoV-2019 genomes. www.Virological.org (2020).

Andrew Rambaut, from the University of Edinburgh, is one of the leaders in phylogenetics. In this blog he has been updating his analysis of the SARS-CoV-2 genomes, the estimated time to the most recent ancestors. The analysis is well commented on and Andrew provides brief reports of recent phylogenetic results. Data used for these analyses are public and can be reproduced by data science aficionados. The same blog presents very interesting studies by other researchers on SARS-CoV-2 and other viruses.

Chapter 5: How Did the COVID-19 Outbreak Start and Evolve?

Kristian, G. Andersen, A. R., Ian Lipkin, W., et al. The proximal origin of SARS-CoV-2. *Nature Medicine* 26: 450–452 (2020).

This is a good perspective offered by some of the leaders in the field that presents the different scenarios for how SARS-CoV-2 could have emerged in humans. It reviews some of the most recent information on the genetic relatives to the new virus, and discusses alternative theories about the recent origin of the virus from bats, pangolins, and other species.

Yan, R., Zhang, Y., Li, Y., et al. Structural basis for the recognition of the SARS-CoV-2 by full-length human ACE2. *Science,* doi:10.1126/science.abb2762 (2020).

This paper shows the detailed structure of the virus binding to the receptor in the cell, the point of entry of the virus to the cell. The structures provide the molecular basis for the development of therapeutics aiming at disrupting this crucial interaction.

van Doremalen, N., Bushmaker, T., Morris, D., et al. Aerosol and surface stability of SARS-CoV-2 as compared with SARS-CoV-1. *New England Journal of Medicine,* doi:10.1056/NEJMc2004973 (2020).

This is a very important study evaluating the lifetimes of the SARS-CoV-2 virus in aerosols and on different surfaces. The findings suggest the main routes of viral spread between individuals.

World Health Organization, *Report of the WHO–China Joint Mission on Corona-virus Disease 2019 (COVID-19)* (World Health Organization, 2020).

This is the joint World Health Organization–China report on the outbreak in Wuhan. It details measurements of infection rates, morbidity and mortality, age distribution, among many other factors. It offers a very interesting description of the clinical characteristics of patients and containment measures taken in Wuhan, Hubei.

Zhou, F., Yu, T., Du, R., et al. Clinical course and risk factors for mortality of adult inpatients with COVID-19 in Wuhan, China: a retrospective cohort study. *Lancet*, doi: 10.1016/S0140-6736(20)30566-3 (2020).

A comprehensive study of 191 COVID-19 hospitalized patients in Wuhan, China comparing survivors versus non-survivors. Detailed analysis is given of the clinical course of the disease in both groups, together with an analysis of risk factors associated with death. These factors include hypertension, diabetes, and coronary heart disease. The study also finds that the median duration of viral shedding in these patients in 20 days, but could be as much as 37 days.

Channappanavar, R., Fett, C., Mack, M., et al. Sex-based differences in suscepti-bility to severe acute respiratory syndrome coronavirus infection. *Journal of Immunology* 198, 4046–4053, doi:10.4049/jimmunol.1601896 (2017).

This paper studies the sex-based differences in the disease and death rates observed in the SARS outbreak using mouse models of SARS. The results show that male mice are more prone to infection than females and that this bias increases with age. These changes include changes in the immune system. Suppression of estrogen in female mice increased the mortality, indicating that hormonal changes could mitigate the effect of the disease.

Liu, W., Zhang, Q., Chen, J., et al. Detection of Covid-19 in children in early January 2020 in Wuhan, China. *New England Journal of Medicine*, doi:10.1056/NEJMc2003717 (2020).

This correspondence delineates some of the findings in infected children in Wuhan, China. The rates and severity of these infections is lower than in the elderly.

Liu, Y., Yan, L., Xiang, T., et al. Viral dynamics in mild and severe cases of COVID-19. *Lancet Infectious Diseases*, doi:10.1016/S1473-3099(20)30232-2 (2020).

This small note reports the clinical characteristics of mild cases and how viral loads can be used as a marker of severity.

Qifang Bi, Y. W., Shujiang M., Chenfei Y., et al. Epidemiology and transmission of COVID-19 in Shenzhen China: analysis of 391 cases and 1,286 of their close contacts, *MedRxiv* (2020).

An analysis of cases in Shenzhen, China, with detailed information on cases, tracing, and observation of close contacts. The study measures time from symptoms to confirmation, isolation, and clinical care. The study also highlights the importance of careful contact tracing, which allowed reducing the time to the implementation of isolation of infected cases. Household analysis showed that children were as likely to be infected as adults.

Holshue, M. L., DeBolt, C., Lindquist, S., et al. First case of 2019 novel coronavirus in the United States. *New England Journal of Medicine* 382, 929–936, doi:10.1056/NEJMoa2001191 (2020).

Analysis of the first COVID-19 case reported in the USA.

Chapter 6: How Does the COVID-19 Outbreak Compare to the SARS Outbreak in 2003?

Zhong, N. S., Zheng, B., Li, Y., et al. Epidemiology and cause of severe acute respiratory syndrome (SARS) in Guangdong, People's Republic of China, in February, 2003. *Lancet* 362, 1353–1358, doi:10.1016/s0140-6736(03)14630-2 (2003).

Ksiazek, T. G., Erdman, D., Goldsmith, C., et al. A novel coronavirus associated with severe acute respiratory syndrome. *New England Journal of Medicine* 348, 1953–1966, doi:10.1056/NEJMoa030781 (2003).

These two papers report the first cases of SARS originating from Guangdong, China, and the isolation of the virus SARS coronavirus. The genetic analysis showed that the SARS cases were all associated with a single coronavirus that spread to other places in the world in 2003.

Peiris, J. S. M., Lai, S., Poon, L., et al. Coronavirus as a possible cause of severe acute respiratory syndrome. *Lancet* 361, 1319–1325, doi: 10.1016/S0140-6736 (03)13077-2 (2003).

An analysis of 50 SARS patients in five transmission clusters in 2003. The authors report the clinical characteristics of these patients and the factors associated with disease severity.

Guan, Y., Zheng, B., He, Y., et al. Isolation and characterization of viruses related to the SARS coronavirus from animals in southern China. *Science* 302, 276–278, doi:10.1126/science.1087139 (2003).

A study looking for SARS-like viruses in different species. Himalayan palm civets were found in a live-animal market in Guangdong, China, infected with viruses similar to the SARS virus. The study highlights the danger of close human–animal interaction in life-animal markets and a possible source of the SARS coronavirus in 2002.

Wang, M., Yan, M., Xu, H., et al. SARS-CoV infection in a restaurant from palm civet. *Emerging Infectious Diseases* 11, 1860–1865, doi: 10.3201/eid1112.041293 (2005).

After the 2003 SARS outbreak was declared to have ended in July 2003, a few isolated SARS infections were reported in humans. In this paper, two individuals were infected with SARS-like viruses in a restaurant in Guangzhou, China. The restaurant served palm civets and had live specimens in the entrance. This isolated event suggests that close human–animal interaction can lead to these zoonotic events.

Hu, B., Zeng, L.-P., Yang, X.-L., et al. Discovery of a rich gene pool of bat SARS-related coronaviruses provides new insights into the origin of SARS coronavirus. *PLoS Pathogens* 13, e1006698, doi:10.1371/journal.ppat.1006698 (2017).

Extensive sampling of SARS-related coronaviruses in bats across different regions of China shows a highly diverse pool of coronaviruses with a large number of recombinations. The large diversity of viruses, including recombinant forms, found in this study suggests that bats could be the main mixing pool and origin of SARS-like outbreaks in humans.

Centers for Disease, Control & Prevention. Severe acute respiratory syndrome: Singapore, 2003. *Morbidity and Mortality Weekly Report* 52, 405–411 (2003).

Joint study of Singapore and the WHO of SARS cases in Singapore in 2003. The work reports on super-spreaders – individuals that are able to infect many contacts.

de Wit, E., van Doremalen, N., Falzarano, D., & Munster, V. J. SARS and MERS: recent insights into emerging coronaviruses. *Nature Reviews Microbiology* 14, 523–534, doi:10.1038/nrmicro.2016.81 (2016).

This is a very good review revising the transmission and pathogenesis of the two previous outbreaks of SARS-like coronavirus in humans, the Middle East respiratory syndrome coronavirus (MERS-CoV) and the acute respiratory syndrome coronavirus (SARS-CoV). I recommend this review by way of introduction to the knowledge we had on SARS-like viruses before the COVID-19 outbreak.

Kuba, K., Imai, Y., Rao, S., et al. A crucial role of angiotensin converting enzyme 2 (ACE2) in SARS coronavirus-induced lung injury. *Nature Medicine* 11, 875–879, doi:10.1038/nm1267 (2005).

This article provides evidence that the angiotensin-converting enzyme 2 protein (ACE2) is a necessary receptor for the virus and that the binding of the viral spike protein can contribute to the disease by dysregulating the renin–angiotensin pathway. It provides evidence on how SARS could cause lung disease, suggesting potential routes for therapeutic intervention.

Zou, L., Ruan, F., Huang, M., et al. SARS-CoV-2 viral load in upper respiratory specimens of infected patients. *New England Journal of Medicine*, doi:10.1056/NEJMc2001737 (2020).

The researchers report the results of SARS-CoV-2 close monitoring of 18 patients in Guangdong, China. The study includes family clusters and an asymptomatic case. The viral load in the asymptomatic patient was similar to the loads in symptomatic patients, indicating the possibility of transmission through asymptomatic or minimally symptomatic patients. The results suggest that the containment strategies for COVID-19 need to be different to ones used to successfully eradicate SARS in 2003.

Chapter 7: How Does the COVID-19 Outbreak Compare to Seasonal and Pandemic Influenza?

Johnson, N. P. & Mueller, J. Updating the accounts: global mortality of the 1918–1920 "Spanish" influenza pandemic. *Bulletin of the History of Medicine* 76, 105–115, doi:10.1353/bhm.2002.0022 (2002).

This is a historical account that re-examines some of the numbers of the deceased in the influenza pandemic of 1918. The analysis suggests the number of worldwide deaths to be of the order of 50 million, and probably higher. Subsequent analyses have challenged some of these estimates.

Grist, N. R. Pandemic influenza 1918. *BMJ* 2, 1632–1633, doi:10.1136/bmj.2 .6205.1632 (1979).

A letter written on September 29, 1918 narrating the devastating effects of the 1918 influenza in a military camp in Massachusetts.

Bootsma, M. C. & Ferguson, N. M. The effect of public health measures on the 1918 influenza pandemic in U.S. cities. *Proceedings of the National Academy of Sciences* 104, 7588–7593, doi:10.1073/pnas.0611071104 (2007).

This is a good study comparing the effects of the 1918 influenza pandemic across several cities in the USA. The study shows that the scope, extent, and the initiation time of public health measures have strong effects on mortality. Early interventions significantly reduced mortality. Some cities were able to significantly reduce transmission rates with early and effective intervention measures.

Chapter 8: Are There Therapeutic Options?

Chu, C. M., Cheng, V., Hung, I., et al. Role of lopinavir/ritonavir in the treatment of SARS: initial virological and clinical findings. *Thorax* 59, 252–256, doi:10.1136/thorax.2003.012658 (2004).

A clinical study in 2003 for 41 SARS patients receiving a combination of lopinavir/ritonavir and ribavirin, compared with 111 patients treated with ribavirin. This small trial showed some encouraging results.

Cao, B., Wang, Y., Wen, D., et al. A trial of lopinavir–ritonavir in adults hospitalized with severe Covid-19. *New England Journal of Medicine*, doi:10.1056/NEJMoa2001282 (2020).

A 199 COVID-19 hospitalized patient study comparing lopinavir–ritonavir to the standard care carried out in Wuhan, China. There was no reported benefit in mortality of reduction of viral load between the two groups.

Weingartl, H., Czub, M., Czub, S., et al. Immunization with modified vaccinia virus Ankara-based recombinant vaccine against severe acute respiratory syndrome is associated with enhanced hepatitis in ferrets. *Journal of Virology* 78, 12672–12676, doi:10.1128/JVI.78.22.12672-12676.2004 (2004).

A study of a SARS vaccine carried out in ferrets in 2004. It shows that the SARS vaccine elicited a strong neutralizing antibody response. However, a strong inflammatory response leading to liver damage was common in infected immunized animals. This study underscores some of the potential challenges of vaccines for SARS-like coronaviruses.

World Health Organization, *Managing Epidemics* (World Health Organization, 2018).

A nice and complete report by the WHO explaining the continuing threats of emerging infectious disease outbreaks. It revises the recent history, provides statistics, details medical and public healthcare measures of containment. It highlights 15 deadly diseases with precise information on each one.

Figure and Quotation Credits

Figure 2.1	Reproduced with permission of the World Health Organization
Figure 2.6	Data from surveillance system of daily deaths in Spain, Spanish Center for Epidemiology, Instituto Carlos III: www.isciii.es/QueHacemos/Servicios/VigilanciaSaludPublicaR ENAVE/EnfermedadesTransmisibles/MoMo/Paginas/default.aspx
Figure 2.7(b)	Anadolu Agency / Contributor / Getty Images.
Figure 3.1	Narvikk/istock/Getty Images Plus
Figure 3.3	Reproduced from Wu, F. *et al.* A new coronavirus associated with human respiratory disease in China (2020): https://doi.org/10.1038/s41586-020-2008-3. Reprinted from *Nature* with permission of Springer Nature
Figure 3.5	Courtesy of John Nicholls, Leo Poon, and Malik Peiris/The University of Hong Kong
Figure 4.2	Reproduced from: *Topological Data Analysis for Genomics and Evolution: Topology in Biology* ©Raul Rabadan and Andrew J. Blumberg/Cambridge University Press 2020
Figure 4.4	Reproduced from Patiño-Galindo *et al.* Recombination and convergent evolution led to the emergence of 2019 Wuhan coronavirus. https://doi.org/10.1101/2020.02.10.942748.
Figure 4.5	Reproduced from: *Topological Data Analysis for Genomics and Evolution: Topology in Biology* ©Raul Rabadan and Andrew J. Blumberg/Cambridge University Press 2020

Figure 4.6   Adapted from Zairis *et al.*, Genomic data analysis in tree spaces: *arXiv: 1607.07503* [q-bio.GN]

Figure 5.1   Reproduced from Guan *et al.* Clinical characteristics of coronavirus disease 2019 in China (2020): DOI: 10.1056/NEJMoa2002032. Reprinted from the *New England Journal of Medicine* with permission from Massachusetts Medical Society

Figure 5.2(a)   Courtesy of Gareth Jones

Figure 5.2(b)   Reproduced with permission from the International Union for Conservation of Nature. *Rhinolophus affinis*. The IUCN Red List of Threatened Species. Version 2020-1. www.iucnredlist.org. Downloaded on 6 April 2020.

Figure 5.3   Reproduced from Lu *et al.*, Genomic characterisation and epidemiology of 2019 novel coronavirus: implications for virus origins and receptor binding, P565–574 (2020): https://doi.org/10.1016/S0140-6736(20)30251-8.
Reprinted from *The Lancet* with permission from Elsevier

Figure 5.4   Reproduced from Patiño-Galindo *et al.* Recombination and convergent evolution led to the emergence of 2019 Wuhan coronavirus (2020): https://doi.org/10.1101/2020.02.10.942748.

Figure 5.6   Reproduced from Renhong Yan *et al.* Structural basis for the recognition of SARS-CoV-2 by full-length human ACE2 (2020): doi: 10.1126/science.abb2762. Reprinted from *Science* with permission from AAAS

Figure 5.7   Reproduced from Zhou *et al.* Clinical course and risk factors for mortality of adult inpatients with COVID-19 in Wuhan, China (2020): https://doi.org/10.1016/S0140-6736(20)30566-3. Reprinted from *The Lancet* with permission from Elsevier

Figure 5.8   Data from Centro Nacional de Epidemiología, Spain. © Raul Rabadan

Figure 5.9   Reproduced with permission from GISAID (www.gisaid.org)

Figure 6.1   Reproduced from Ksiazek, T. G. *et al.* A novel coronavirus associated with severe acute respiratory syndrome (2003): doi:10.1056/NEJMoa030781 (2003). Reprinted from the

	New England Journal of Medicine with permission from Massachusetts Medical Society
Figure 6.2	Reproduced from de Wit *et al. SARS and MERS: recent insights into emerging coronaviruses* (2016): https://doi.org/10.1038/nrmicro.2016.81. Reprinted from *Nature Reviews Microbiology* with permission from Springer Nature.
Figures 7.1–7.4	Reproduced from: *Topological Data Analysis for Genomics and Evolution: Topology in Biology* ©Raul Rabadan and Andrew J. Blumberg/Cambridge University Press 2020
Figure 7.5	"Emergency hospital during influenza epidemic, Camp Funston, Kansas" (NCP 001603). OHA 250: New Contributed Photographs Collection. Reproduced with permission from Otis Historical Archives, National Museum of Health and Medicine
Figure 7.6	Reproduced from Nelson, M., Holmes, E. The evolution of epidemic influenza (2007): https://doi.org/10.1038/nrg2053. Reprinted from *Nature Reviews Genetics* with permission from Springer Nature.

All other figures are © Raul Rabadan/Cambridge University Press (2020)

Credits for Quotations

Chapter 3	Taken from *The Lives of a Cell: Notes of a Biology Watcher* by Lewis Thomas. Reproduced with kind permission of his estate.
Chapter 4	Taken from *Sydney Brenner's 10-on-10: The Chronicles of Evolution* © 2019 Agency for Science, Technology and Research. Reproduced with permission of Wildtype Media Group and the Agency for Science, Technology and Research.
Chapter 6	Taken from Zhong, N. S. *et al.* Epidemiology and cause of severe acute respiratory syndrome (SARS) in Guangdong, People's Republic of China (2003). Lancet 362, 1353-1358, doi:10.1016/s0140-6736(03)14630-2 (2003). Reproduced with permission from Elsevier.

Index